Molecular Chemistry of Sol-Gel Derived Nanomaterials

Molecular Chemistry of Sol-Gel Derived Nanomaterials

Robert Corriu,
Université Montpellier II, France

Nguyên Trong Anh,
École Polytechnique, CNRS, France

WILEY
A John Wiley and Sons, Ltd., Publication

Originally published in French by École Polytechnique, © Robert Corriu and Nguyên Trong Anh, 2008 (978-2-730-21413-1).

Registered office
John Wiley & Sons Ltd, The Atrium, Southern Gate, Chichester, West Sussex, PO19 8SQ, United Kingdom.

For details of our global editorial offices, for customer services and for information about how to apply for permission to reuse the copyright material in this book please see our website at www.wiley.com.

Library of Congress Cataloging-in-Publication Data

Corriu, Robert.
Molecular chemistry of sol-gel derived nanomaterials/Robert Corriu, Nguyên Trong Anh.
p. cm.
"Originally published in French by École Polytechnique."
Includes bibliographical references and index.
ISBN 978-0-470-72117-9
1. Colloids. 2. Nanofluids. 3. Nanochemistry. 4. Nanostructured materials. I. Nguyên, Trong Anh, 1935- II. Title.
TA418.9.N35. C68313 2009
620.1′1–dc22 2008047063

A catalogue record for this book is available from the British Library.

Typeset in 10.5/13pt Sabon by Thomson Digital, Noida, India.
Printed and bound in Great Britain by TJ International Ltd, Padstow, Cornwall

ISBN 978-0-470-72117-9 (HB)

Contents

Preface

The writing of this book was motivated by the ever increasing interest in the rapid development of nanosciences and nanotechnologies. As scientists in the field, we are perturbed that nanosciences are *de facto* perceived as physics. Admittedly, the "nanoworld" is studied with physical instruments (e. g. scanning, tunneling and atomic force microscopes) and these studies are important, as it is already known that physical properties vary at different scales. Also, nanotechnologies have precipitated a miniaturization race, especially in electronics, following the famous aphorism 'There is plenty of room at the bottom' (R. P. Feynman). This miniaturization is essentially carried out by physical methods. This has led to nanosciences and nanotechnologies being identified by a large part of the general and the scientific community as a new physical domain, and therefore as no concern to chemistry.

It seemed necessary to us to amend this point of view by outlining the possibilities opened up by chemistry in this very promising field. Let us remember that nanosciences study *nano-objects* (entities of nanometric sizes) and their assembling into *nanomaterials*. Chemists have always thought in terms of nanometric objects (atoms, ions, molecules, etc.). Chemical syntheses are a planned assembling of these elementary units. Thus the 'bottom-up' approach in nanosciences is simply an application of familiar chemical ways of thinking and doing in a new domain.

Chemistry has also become in the recent past a creative science. To assert that chemists, with the tools already available, can prepare any conceivable structure is neither false nor extravagant. Therefore, the 'know how' of molecular chemists in synthetics can play a significant role in nanosciences. This is presented in Chapters 1 and 2 with particular emphasis on the potential development of new materials exhibiting specific physical or chemical properties.

The focus of this book is on the new possibilities in material science opened up by the recent advances in inorganic polymerizations, better known as sol-gel processes. These ancient methods[1] sank into oblivion and were not rediscovered until the 1950s when chemists in the glass industry took advantage of the passage through a viscous state in order to shape the glasses and/or to transform them into coatings (see Chapter 3). Even then, for many years the primary concern was with industrial problems; only in the last twenty years have fundamental studies been undertaken in order to exploit the potential of these methods more widely.

Sol-gel processes are inorganic polymerizations which obey similar although more complex rules to organic polymerizations. The solid state chemistry approach produces two major new routes to original materials. On the one hand, there are the 'chimie douce' (or 'mild chemistry') methods[2] which allow complete compatibility between organic or biological and inorganic components. On the other hand, there are these sol-gel processes which lead to new materials through kinetically controlled syntheses, a usable complementary alternative to the customary thermodynamically controlled syntheses. If we recall, traditional preparations of glasses and ceramics take place at high temperatures (>400 °C and very often in the 1000–2000 °C range) which usually destroy organic and biological molecules. Thus, what started as a simple improvement to industrial processes has become a *bona fide* revolution which drastically changes inorganic synthesis.

We can now prepare materials which were previously unfeasible; it is already possible to obtain solids in which organic, organometallic or even biological entities can be incorporated or chemically bonded to inorganic matrices. This could open up a whole new field of chemistry to be explored, as the majority of materials obtained up to now are silicon hybrids, due to the ability of silicon to bind to carbon and to sustain controlled polymerization. We have not yet mastered the polymerization of other oxides (SnO_2, TiO_2, Al_2O_3, NiO, etc.), in order to take advantage of their semiconducting (SnO_2), photovoltaic (TiO_2) or magnetic (Fe_3O_4) properties (properties that do not exist in SnO_2), nor do we know how to combine them with organic molecules. Likewise, nitride and phosphide matrices have not been studied yet. In time, these hybrid organic-inorganic materials could become an inexhaustible source of new materials. In Chapters 3–6 we show some initial results which describe the present state of the art.

Although many types of hybrids are described in this book, the emphasis is on nanostructured materials (Chapter 5). These materials correspond to the polymerization of organic entities having at least two carbon atoms substituted by $-Si(OR)_3$ groups. The hydrolytic (sol-gel) polycondensation of these precursors provides materials in which the organic components are inseparable from the inorganic parts. The organic entities are evenly ordered in these solids, which for this reason are called nanostructured hybrids. In fact, two independent organizations are observed: a nanometric structure revealed by X-ray diffraction and a micrometric structure confirmed by white light birefringence. The nanometric organization is generated in the colloidal sol phase and the micrometric organization during the ageing of the solid gel phase. This type of non-crystalline organization, never observed before, shows another interesting feature of inorganic polymerization. It has been detected with precursors having linear, planar, twisted or tetrahedral geometries.

Chapter 6 describes some developments of the mesoporous materials discovered by Kresge *et al.*[3] For the first time in the history of molecular and macromolecular chemistry, it has become possible to precisely locate the relative positions of different chemical entities. When these entities have distinct properties, materials can be obtained in which interacting properties occur at the nanoscale (until now such interactivity has only been observed between supramicrometric layers). Indeed, molecules, organometallic or coordination complexes, and metallic or inorganic (oxides, nitrides, phosphides) nanoparticles can add different physical (magnetic, optical, electrical, etc.) or chemical (catalysis, sequestration, separation, etc.) properties.

Some potential developments of the chemistry of hybrids will be presented in Chapter 7, along with some present and future applications. The molecular approach can fully exploit the synthetic capacities of chemistry and in contrast to the traditional one-step thermodynamically controlled preparations; sol-gel processes involve several kinetically controlled stages and allow the marriage of molecular chemistry and material science.

Nanosciences are defined purely by the size of the objects studied. This is best served by the cooperation of various parties employing different and complementary competencies in order to produce a multidisciplinary approach. Exploiting the materials' properties requires teamwork of various expertises, ranging from molecular synthesis through physical chemistry to technologies. New materials will result from this interdisciplinary synergy and their preparation necessitates a diverse knowledge base. Nanomaterials with interacting properties are no longer an ideal (see Chapter 6).

Robert Corriu,
Université Montpellier II, France

Nguyên Trong Anh,
CNRS, France

References

[1] (a) M. Ebelmann, *Ann. Chim. Phys.* **1846**, *16*, 129; (b) M. Ebelmann, *C. R. Acad. Sci. Paris* **1847**, *25*, 854.
[2] (a) J. Livage, *Chem. Scr.* **1988**, *28*, 9; (b) J. Rouxel, *Chem. Scr.* **1988**, *28*, 33.
[3] C.T. Kresge, M. E. Leonowicz, W. J. Roth, J. C. Vertuli, J. S. Beck, *Nature* **1992**, *359*, 710.

About the Authors

Robert Corriu

Robert Corriu is Emeritus Professor at the University of Montpellier II (France). He is a member of both the French Academy of Sciences and the French Academy of Technologies. He is the recipient of three major international scientific prizes in the silicon field: Wacker Silicon Award (1998), Humboldt Research Award (1992) and the ACS Kipping Award (1984) and has also received accolades for his work from Japan and Germany. Working in the fields of organometallic chemistry and the organometallic chemistry of silicon, Professor Corriu is particularly interested in the opportunities that molecular chemistry can create in the field of materials science using the sol-gel process.

Nguyên Trong Anh

Nguyên Trong Anh was formerly Director of Research at the Centre National de la Recherche Scientifique, Professor of Chemistry at the Ecole Polytechnique (Palaiseau, France) and Editor in Chief of *New Journal of Chemistry.* Trained as an experimental organic chemist, he became interested in applied theoretical chemistry and has worked on problems of organic stereochemistry and reaction mechanisms. He is the author of several books including "*Frontier Orbitals*" and "*Les Règles de Woodward-Hoffmann*" first published in 1971 which has since been translated into German, Italian, Spanish and Japanese.

1

Molecular Chemistry and Nanosciences

1.1 INTRODUCTION

Nanosciences study *nano-objects*, *i.e.* nanometric-size objects ($1\,\mathrm{nm} = 1 \times 10^{-9}\,\mathrm{m}$) and their transformation into *nanomaterials*.[*†] Unquestionably, they represent a most promising field of material sciences for the next few years. The main challenge will be the control of physical and chemical properties by methods operating at atomic or molecular level.

However, in the mind of many scientists, physics is the major factor in nanosciences, chemistry playing but a minor role. This opinion is largely the consequence of the historical development of nanosciences, as explained in the next section.

The purpose of this book is to amend this view by pointing out the potential of chemistry in this area. We shall present in Section 1.2 the two principal approaches in nanosciences (the 'top-down' approach which relies mostly on physics and the 'bottom-up' approach which is essentially a matter of chemistry), and relate in Section 1.3 how chemistry has evolved from an exploratory to a creative science. Chemistry can now tackle successfully a great variety of problems, from the creation of new

---

*Thus nanosciences are defined by the size of the objects, rather than by the nature of the phenomenon studied as in optics, electricity, etc. It follows that they are by definition multidisciplinary.

†Nanomaterials differ from 'ordinary' materials in that their properties can be traced back to those of their nano-object component: in other words, these properties are already incorporated at the nanoscale.

Molecular Chemistry of Sol-Gel Derived Nanomaterials Robert Corriu and Nguyên Trong Anh

materials to the synthesis of auto-organized systems which can almost mimic living matter. With the synthetic methods already perfected and/or to be discovered in the near future, chemistry can convert nano-objects into a vast number of operational materials, exemplified by carbon and ceramic fibers, the forerunners of nanomaterials (Section 1.4).

Nanosciences are multidisciplinary, with physics and chemistry as natural partners. Chemistry can create new molecules, particles, nano-objects, etc., which can lead to innovative designs for new materials, *e.g.* materials in which several physical or chemical properties interact. If their preparation is the chemist's responsibility, the study and utilization of these materials' original properties come under the remit of the physicist. Other disciplines may be involved as well. For instance, mechanics will be implicated because no materials exist without mechanical properties. Mechanical attributes can also be fine-tuned at the nanometric scale. Biology is less directly involved because most biological entities exceed a micrometer in size; however, it will benefit from the development of nano-objects capable of working in a biological environment. The most illustrative example is that of biosensors capable of detecting and measuring certain substances *in situ* (e.g. in blood). Furthermore, modeling biological properties may suggest new designs for nanomaterials. Thus the membrane phospholipids have served as a model for the development of vesicle-forming surfactant compounds.

Modern science is demanding, requiring expert knowledge from each contributing discipline. Only close cooperation between experienced and competent specialists, who are able to communicate with each other, understand each other and conceive a joint project, can lead to new and significant achievements.

1.2 SCOPE AND ORIGIN OF NANOSCIENCES: THE 'TOP-DOWN' AND 'BOTTOM-UP' APPROACHES

Although chemists handle objects of nanometric sizes daily, physicists must be credited with formalizing the concept of nanosciences. This is due to two reasons.

The first one is purely scientific. It comes from the quasi-certitude that exploration of the 'nanoworld',[1] that is to say matter at the nanometer scale, will lead to the discovery of new, unexpected physical properties. Indeed, it is known that physical properties are dependent on the observation scale: studies at the micrometric scale will not reveal the same properties as studies at the nanometric scale. Investigation of the behavior of isolated units (metal atoms, particles, molecules) becomes possible with the invention of the atomic force microscope and the scanning tunneling microscope. Some results obtained are spectacular and open up exciting vistas to scientists. For example, physicists have been able to study the transition of a single electron from the fundamental to the excited state in semiconductors as well as in suitably chosen organic molecules. IBM scientists have written their company's acronym on an appropriate surface by displacing atoms one by one. To recap, physics has the instruments for exploring the nanoworld and the capacity to study and exploit the (optical, electrical, magnetic, etc...) properties of nano-objects.

The second reason, more technological, has economic motivations. The mass diffusion of electronic products and their involvement in almost everyday activity have generated a mounting need for smaller and yet more powerful microprocessors. This demand is quantified by the famous Moore law which predicts that the performance of electronic components increase by one order of magnitude every two years. Microprocessors are, therefore, miniaturized and tend towards 'nanoprocessors'. This approach has been termed 'top-down' and corresponds to the first manifestation of the nanoscience concept. From an economic point of view, the top-down methodology is unquestionably the most important approach at present and has created a lively international competition.

There is also a symmetrical approach called 'bottom-up' in which the nanomaterial is chemically assembled from elementary chemical components, just like a wall is constructed from bricks and mortar. While the top-down approach is essentially a miniaturization technology from which chemistry is absent, the bottom-up approach, based on synthesis, fits perfectly with chemical methodology. The building blocks – molecules, molecular complexes, atoms or aggregates, all entities whose sizes vary from tenths of a nanometer to tens of nanometers – are familiar to chemists. The assembling methods (the mason's mortar) use inclusion and polymerizations of organic or inorganic entities. As shall be explained in the next chapter, chemistry possesses all the necessary requirements for developing nanosciences by the bottom-up approach.

One of the most illuminating examples concerns the *selective elimination* of lead from drinking water.[2–4] After passage through a filtering cartridge, the Pb^{2+} concentration is $<5\ \mu g\, l^{-1}$. The concentration of other ions (Na^+, Ca^{2+}, Mg^{2+}, etc.) is unchanged. This achievement, unbelievable just 10 years ago, is now possible because coordination chemists can prepare compounds capable of chelating selectively different metal ions. These compounds are incorporated into solids by polymerizations. In this case, a Pb^{2+}-selective chelating molecule was bonded to silica, resulting in a material, which can be shaped into cartridges. This example is proof that chemistry can synthesize operational and selective nanomaterials.

However, physics is not absent from the bottom-up approach. Some nano-objects, for example fullerenes and carbon nanotubes, can only be obtained by physical methods. There exist also physical assembling methods: vapor phase deposition, molecular beam, etc. All these approaches can lead to new materials.

1.3 CHEMICAL MUTATION: FROM AN EXPLORATORY TO A CREATIVE SCIENCE

During the last fifty years, science has progressively metamorphosed.‡ Let us illustrate these changes with some examples, with particular emphasis on synthesis, which is the foundation of chemical creativity.

A revolution in structural determination launched this chemical mutation. In the late 1950s, recording spectrographs gradually allowed chemists to complete chemical analyses with physical methods (IR, UV, NMR, EPR, MS, X-ray diffraction, etc.) An exhaustive list would take too long and be too difficult to provide, with the number of these identification methods being very large and increasing by the day. Note, however, that chemical quantitative analysis remains *a necessary safeguard* in material sciences (we shall return to this point in Chapter 6).

These analytical tools have permitted a better comprehension of reactivity. Mastering the concepts governing the formation of chemical entities, the organization of solids and molecular structure has enabled chemists to synthesize incredibly complex molecules. Thus, Professor Y. Kishi's group has prepared palytoxin, a natural product isolated from soft coral. This compound[5,6] possesses 62 chiral carbons and has 2^{62} ($\sim 4 \times 10^{18}$) stereoisomers (Figure 1.1). On account of the precision of existing synthetic methods, it has been possible to produce the natural isomer.

‡At the end of the nineteeth century, classical physics was a coherent corpus of doctrines, able to rationalize practically all known phenomena, thanks to mechanics, thermodynamics and electromagnetism. As for chemistry, which was largely empirical during the nineteenth century, it had sufficiently progressed by the middle of the twentieth century to be considered as 'having come of age'. Indeed, fundamental concepts like covalent bond or aromaticity, initially introduced empirically, can be explained by quantum mechanics. Students no longer need to learn by rote hundreds of reactions; they have only to understand a dozen mechanisms (additions, eliminations, substitutions, rearrangements, etc.) Also, the number of complex multistage syntheses already realized show that organic chemists could synthetize practically any existing molecule.

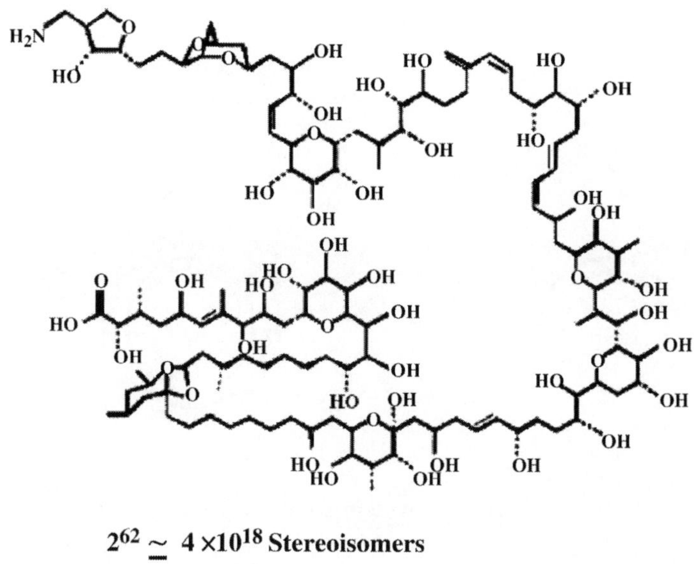

Figure 1.1 Palytoxin. Reproduced by permission of *L'actualité chimique*

Other exotic molecules have also been made. One instance is cubane[7] (Figure 1.2) whose carbon atoms have valence angles of 90° instead of 109° 28′. Figure 1.2 also shows a tetrahedral polymetallic cluster which can be resolved into its two optical isomers[8] and a scale polymer in which carbon atoms have been replaced by silicon atoms.[9] Very different elements can now be bonded together and the size of polymetallic clusters controlled.[10,11] An outstanding achievement of inorganic chemistry is the synthesis of superconductor ceramics (YBaCuO).[12,13]

These – far from exhaustive – examples demonstrate that *chemists can now synthesize any imaginable structure*. Chemistry has left for good the exploratory domain to become a science of creation.

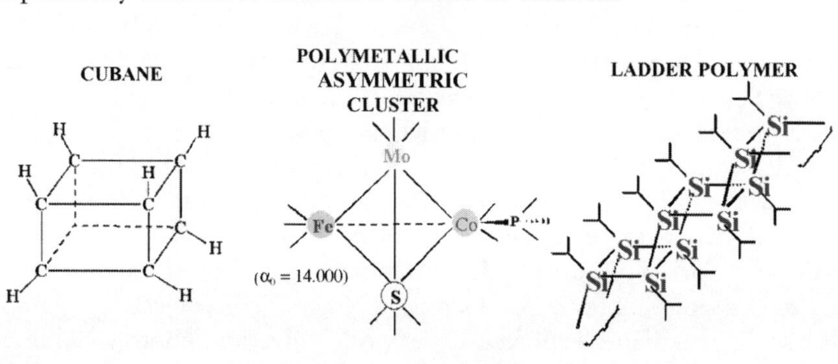

Figure 1.2 Some unusual compounds which have been synthesized. Reproduced by permission of *L'actualité chimique*

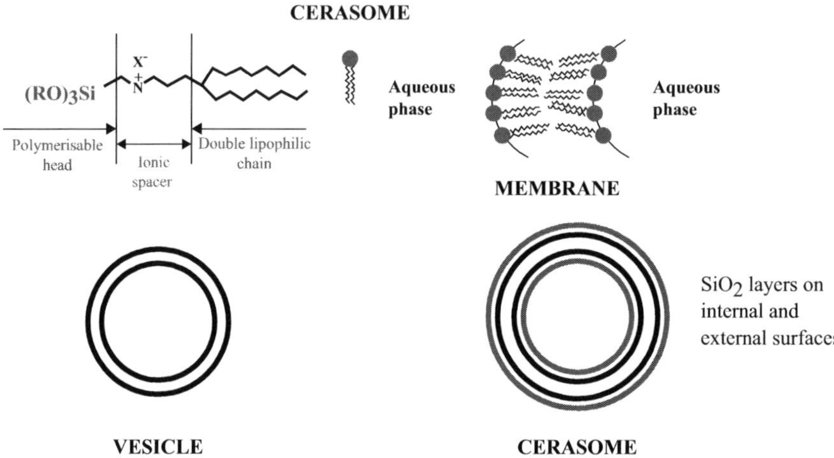

Figure 1.3 Preparation of vesicles with their internal and external surfaces protected by a film of SiO_2

Let us mention also the recently published example of the cerasomes.[14] These vesicles have been obtained using specific chemical methods mimicking the formation of biological membranes. The membrane of these vesicles is coated externally and internally with a molecular layer of silica, giving rise to well-defined systems, which, however, can undergo controlled exchanges (Figure 1.3).

This example shows that chemistry can synthesize not only *new structures* but also structures with *novel properties*, for physical properties can now be correlated with chemical structures. Since the 1970s, material sciences, particularly the chemistry of inorganic solids, have extensively studied the physical properties of chemical products. Later on, macromolecular and molecular chemistries successfully prepared, from organic or organometallic building blocks, materials having special physical properties. Organic conductors were a historical watershed: for the first time, molecular systems prepared by methods of organic synthesis showed conducting or even superconducting properties, which until then were specific of metals.[15,16] Things fell into place when scientists realized that delocalized π electrons in unsaturated organic molecules are comparable with the electrons responsible for metal conduction and can therefore induce the same properties. Subsequently, several other types of polymers with various physical properties have been discovered.[17] Figure 1.4 presents some conducting polymers, a piezoelectric polymer and some polysilanes endowed with (semiconduction, photo-oxidation, thermochromism) properties related to the Si–Si σ bond.[18] Study of these properties has led to a most interesting theoretical development.

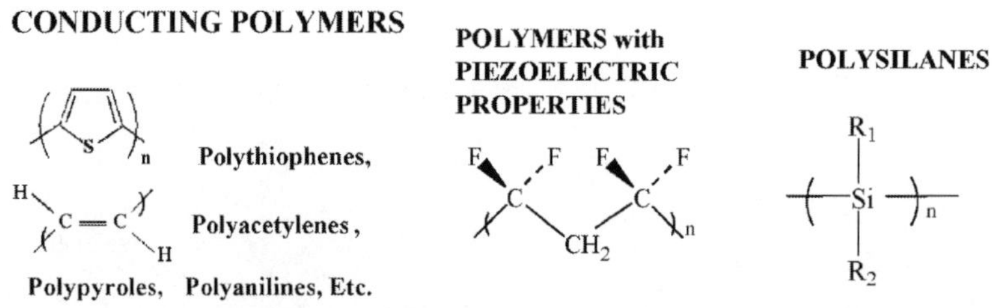

Figure 1.4 Examples of polymers with remarkable physical properties. Reproduced by permission of *L'actualité chimique*

The outstanding optical properties of lanthanides are another significant example. They are responsible for color TV, for signal transmission by optic fibers as well as for remarkable photoluminescent properties. Let us also draw attention to the nonlinear optical (NLO) properties of new organic molecules (Figure 1.5). Nonlinear optics is the branch of optics, which describes the behavior of light in nonlinear media, in which the polarization responds nonlinearly to the electric field of the light. This nonlinearity is only observed at very high light intensities, such as those provided by pulsed lasers.

In the last few years, chemistry has succeeded in creating new entities having expected or unexpected properties. Here are two examples.

The first example is a new technology for generating metallic nanoparticles in mild conditions. This discovery by Bruno Chaudret[19] advantageously replaces the preparation of nanoparticles by reduction of metallic salts. It is based on the very mild decomposition of coordination complexes in which the metal is feebly chelated (π complexes). The growth of the nanoparticle is controlled by weakly coordinating additives, which limit the growth while protecting the metallic entities (Figure 1.6).

The second example comes from dendrimer chemistry (dendrimers being molecules replicating in space from a center, like a cauliflower). The different branches are identical and the size of such molecules can be quite large (Figure 1.7). Phosphorus compounds are very convenient

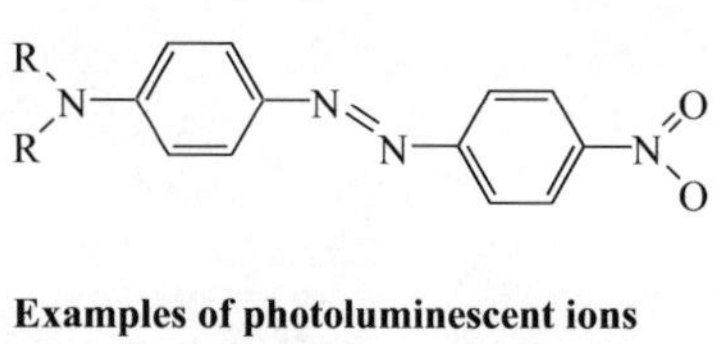

Examples of photoluminescent ions

Tb^{3+} green, **Eu^{3+}** red, **Er^{3+}** infrared

Figure 1.5 Molecule with NLO properties used for frequency doubling. Reproduced by permission of *L'actualité chimique*

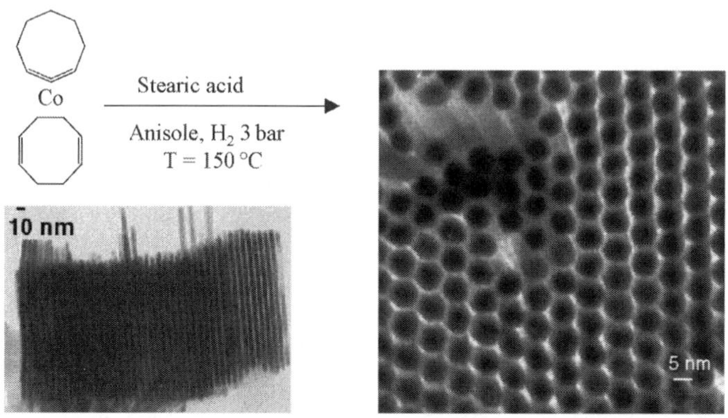

Figure 1.6 Nanorod superlattice of Co nanoparticles obtained by controlled decomposition of a Co π complex. Reproduced with permission from *Angewandte Chemie International Edition*, Unprecedented crystalline super-lattices of monodisperse cobalt nanorods by Dumestre, Frederic; Chaudret, Bruno; Amiens, Catherine; Respaud, Marc; Fejes, Peter; Renaud, Philippe; Zurcher, Peter, *42*, 5213–5216. Copyright (2003) Wiley-VCH

as they permit regular growth of the dendrimers and can be analyzed by ^{31}P NMR.[20] From a close collaboration with biologists and physicians, Jean-Pierre Majoral and his group have been able to establish that phosphorus dendrimers show unexpected therapeutic properties.[21]

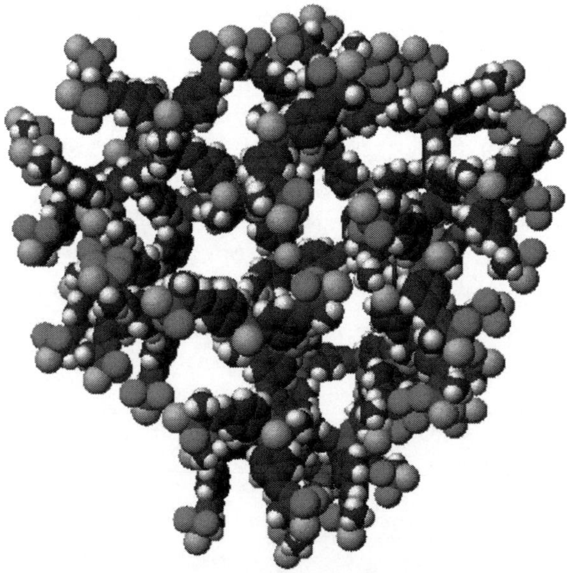

Figure 1.7 Molecular model of a phosphorus dendrimer presenting 48 $P(S)Cl_2$ groups on its surface (See Plate 1 for color representation)

They remarkably increase immune defenses against cancerous cells by developing 'Natural Killers' (NK), which are the equivalent of white blood cells, capable of phagocyting the malignant cells. This totally unpredictable discovery illustrates how many surprises the creative power of chemistry can have in store. In the present case, the biological mechanisms, which induce the growth of NK, are completely unknown.

1.4 CARBON AND CERAMIC FIBERS: THE NANOMATERIAL 'ANCESTORS'

Carbon and ceramic fibers meet the definition of nanomaterials. These compounds with remarkable mechanical properties have been prepared from a *single molecular precursor*, assembled and shaped in the course of a series of chemically controlled steps. Carbon fibers were prepared in the 1960s, with ceramic fibers being prepared in 1975. These materials cannot be obtained by classical thermal methods. In both cases, innovative approaches have opened up new horizons of research. However, only carbon fibers, of low production costs and wide applications, have been a commercial and industrial success. Ceramic fibers have outstanding properties. Unfortunately, their cost has not encouraged industrial production.

1.4.1 Carbon Fibers

We shall now sketch out the preparation of the polyacrylonitrile (PAN) carbon fiber. The story begins in the early 1960s with the discovery of unexpected properties of the solids obtained by pyrolysis of PAN. Figure 1.8 shows schematically the reactions occurring during the successive pyrolyses carried out at temperatures ranging from 200 to 1300 °C under inert atmosphere (He) and various orientation constraints. In the first stage, this polymer is transformed into heterocyclic polycondensates. In the second stage, the hydrogen and nitrogen are eliminated. The solid becomes a nanometric ribbon of polyaromatic units. Under the imposed constraints, these ribbons twist together, becoming entangled into larger fibers, in the manner of jute fibers, which wind up and give a string, then a rope.

These fibers can be spun into great lengths by industrial methods. In composites, they can play the role of the metallic structure in reinforced

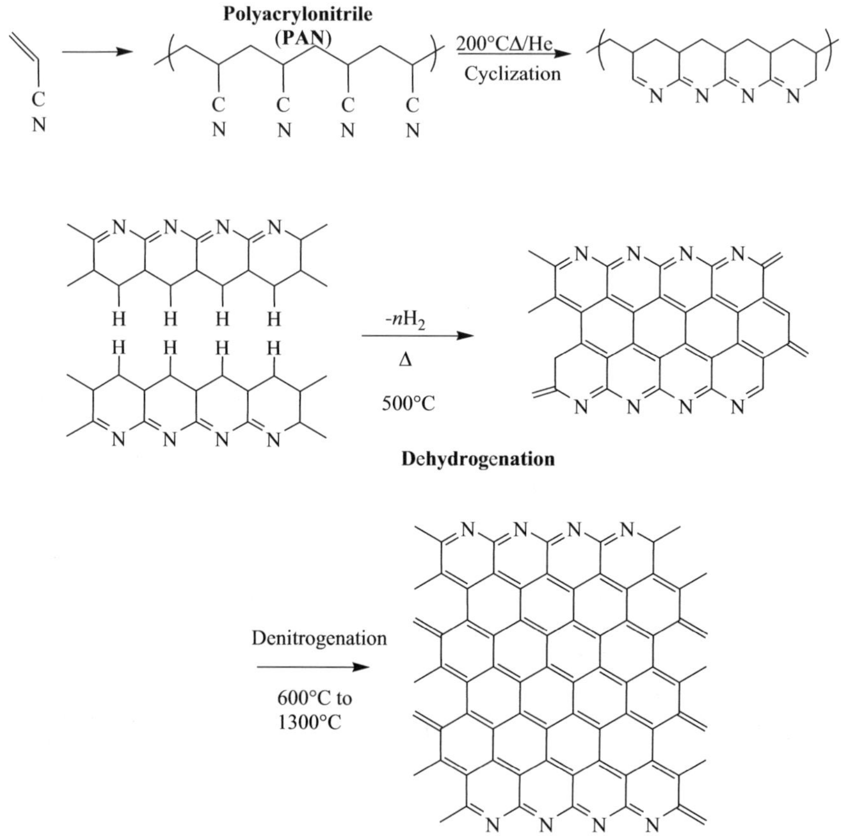

Figure 1.8 Schematic representation of the preparation of carbon fibers from polyacrylonitrile (PAN) by successive polycondensations of carbon atoms

concrete. Their inclusion in a polymer matrix results in materials, which are remarkably resistant to stretching and deformations.

At present, there exist a great number of carbon fibers. Those with weak mechanical properties are used for filtration, thermal isolation, gas adsorption and heat dissipation in braking systems. Those having good mechanical properties are utilized in composites intended for more exacting applications (e.g. in aeronautics). Composites of carbon fibers in appropriate matrices offer, for the same weight, the best resistance to the mechanical constraints experienced by a flying aeroplane. Many carbon fibers, other than PAN fibers, have been prepared from other polymers (polyesters, polyamides, etc.). Each has different characteristics and different uses.

The making of carbon fibers represents an important innovation, not only technologically, but also conceptually: organic textile polymers have been transformed by a simple thermal treatment into materials capable of competing with metals for some applications. In addition, they are easily shaped and are much lighter.

It is perhaps slightly inaccurate to present these fibers as being conceived from the molecular scale, since PAN has been known for a long time. It is clear, however, that the discovery of PAN fibers would not have occurred, had chemists not studied the development of the polymer polycondensation and realized the possibility of obtaining materials with properties radically different from those of the initial polymer. Carbon fibers have served as models for the ceramic fibers to be discussed in the next section.

1.4.2 SiC, Si_3N_4 Ceramic Fibers

If carbon fibers can be considered as remote ancestors of nanomaterials, their 'homo erectus' so to speak, then the SiC and Si_3N_4 fibers are truly the 'first' nanomaterials, their Cro-Magnon ancestors.

The discovery of carbon fibers was the fruit of observation and their industrial success the outcome of perspicacity. Their story represents a beautiful example of serendipity. However, the invention of ceramic fibers was definitely not fortuitous. The conquest of space has created a need for materials with exceptional mechanical properties, capable of resisting temperatures above 1000 °C. Carbon materials, very sensitive to oxidation, cannot meet these requirements, but ceramic materials like SiC and Si_3N_4 can resist high temperatures, even in oxidative media.

Ceramic materials were the first nanomaterials to be prepared. Their final properties had been planned at the molecular scale and their synthesis was carried out step by step from a single molecule. The synthesis was designed to prepare a SiC (or Si_3N_4) ceramic fiber.

Long before the formulation of the nanoscience concept, Verbeek[22] (1974, Germany) and Yajima[23,24] (1975, Japan) independently worked out a method for preparing SiC (Yajima) and Si_3N_4 (Verbeek) ceramic fibers with excellent thermomechanical properties.

We shall deal here with the SiC case, which is much better known than the Si_3N_4 case, since Yajima, having an academic position, has abundantly published, whereas Verbeek, who works in industry, has filed patents. Silicon carbide, a covalent material, is a material highly resilient to mechanical constraints (abrasion, traction or torsion). Its chemical and

$$SiO_2 + 2C \rightarrow 2\,CO_2 + SiC$$

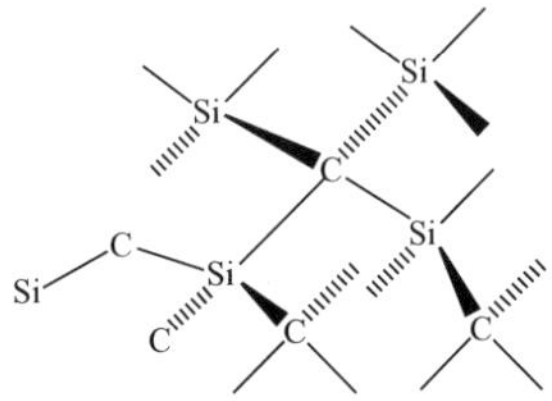

Figure 1.9 SiO_2 carboreduction

mechanical stabilities at high temperatures (~1500 °C) are truly remarkable.

Powdered and solid SiC have been known for a long time. The simplest way to prepare them is by total carboreduction of SiO_2 (Figure 1.9). However solid SiC, being too resistant, cannot be drawn into fiber, molded or coated. Compacting SiC or Si_3N_4 powders is also hopeless. Both Verbeek and Yajima have independently invented a completely new method for preparing ceramics. The guiding principle is the following: the ceramics must pass through an intermediate state *viscous enough to permit the material to be drawn into fibers*. Clearly this stage should precede the final stage of ceramization. These authors have then elaborated a general scheme similar to the different steps of the sol-gel process. The difference is that the sol-gel process proceeds at room temperature whereas the final ceramization, according to Yajima and Verbeek, requires a high temperature (Figure 1.10).

The Yajima process (Figure 1.11) is based on the polymerization of dimethyldichlorosilane, $(CH_3)_2SiCl_2$, a compound produced in great quantities by the silicon industry.

Polymerization of $(CH_3)_2SiCl_2$ in the presence of sodium metal gives a polysilane with Si-Si bonds. Heated at 350 °C, its linear skeleton undergoes the so-called Kumada rearrangement[25] (Figure 1.12).

This rearrangement produces polycarbosilane chains containing the Si–C–Si bonds of the ceramics. The functional Si–H bonds, which are

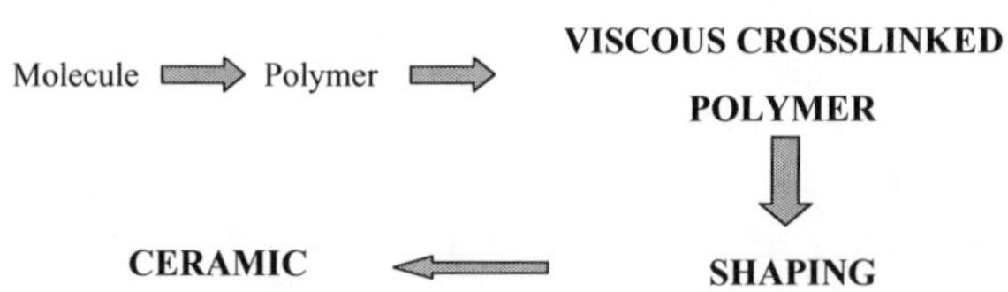

Figure 1.10 Schematic representation of SiC shaping

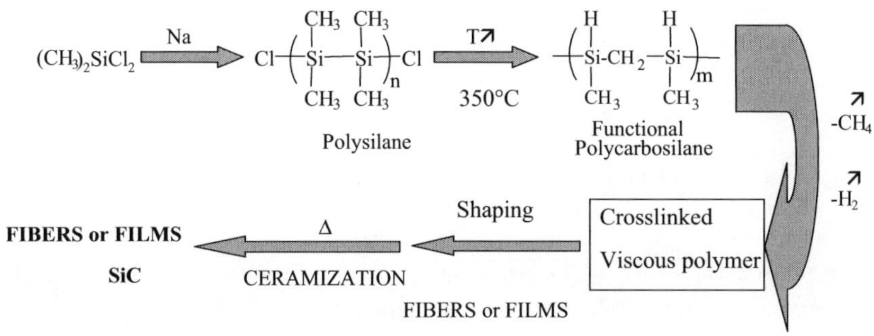

Figure 1.11 Schematic representation of the chemical transformations necessary for SiC shaping. Reproduced with permission from *Journal of Organometallic Chemistry, Organosilicon Chemistry and Nanoscience* by Robert Corriu, *686*, 1–2, 32–41. Copyright (2003) Elsevier

concomitantly made, will permit the reticulation of the linear chains of polycarbosilane, leading to a three-dimensional system. This set of reactions occurs during a controlled pyrolysis which at ~450 °C gives a viscous reticulum, which can be drawn into long fibers or be made into coatings for the protection and reinforcement of other materials. It is also possible to obtain composites with SiC or Si_3N_4 matrices containing reinforcement additives or fibers of another material. The last step is the ceramization of the material with the elimination of residual elements (essentially CH_4 and H_2).

The various stages of this reaction scheme have been optimized in order to increase the fiber yield and minimize the oxycarbides resulting from an ancillary oxidation quite difficult to avoid at the temperatures employed.[26]

The Yajima and Verbeek processes constitute remarkable conceptual progress. The major inconvenience of the Yajima process is the use of an alkaline metal in stoichiometric amounts. We have described a much more handy catalytic preparation of polysilane using a catalytic polymerization of $R_1R_2SiH_2$ into $(R_1R_2Si)_n$ discovered by Harrod and Samuel.[27,28] The reaction in Equation (1.1) shows the direct polymerization of a hydrosilane in very mild conditions.

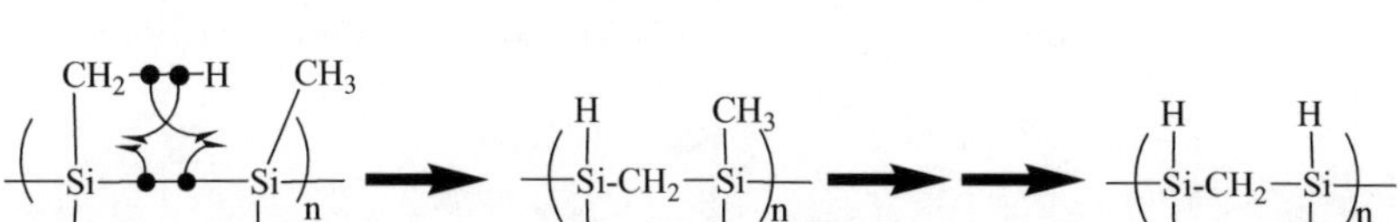

Figure 1.12 Mechanism of the Kumada rearrangement

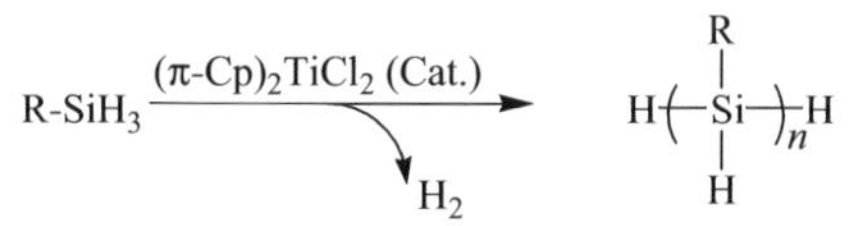

Equation 1.1 Catalytic polymerization of polyhydrosilanes

The polycarbosilane is obtained in one step from pentahydrodisila-1,4-hexane, a molecular precursor which is directly polymerized and polycondensed in solution in the presence of the (π-Cp)$_2$TiMe$_2$ catalyst (Figure 1.13).[29] The proposed mechanism for this reaction involves the reduction of the catalyst by the Si–H bonds, generating a complex of the [Cp$_2$Ti] type. This complex catalyzes the formation of Si–Si bonds by oxidative addition and reductive elimination. The reticulation of the polycondensate (and by way of consequence, its viscosity) can be regulated by a carefully controlled admission of air, which provokes an oxidative poisoning of the catalyst. When the crosslinking of the chains reaches a suitable viscosity, one can proceed to the shaping and then the ceramization of the material. The poisoned catalyst is steadily regenerated *in situ* by reaction with the reducing Si–H bonds. This example nicely illustrates the flexibility of molecular chemistry: preparation of ceramic fibers or coatings is made in one single catalytic step leading directly to the easily shaped viscous material.

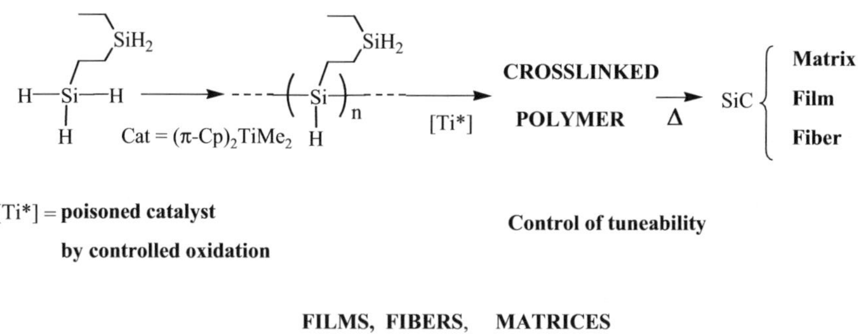

Figure 1.13 Catalytic process for the control of tuneability of SiC. Reproduced with permission from *Journal of Organometallic Chemistry, Organosilicon Chemistry and Nanoscience* by Robert Corriu, *686*, 1–2, 32–41. Copyright (2003) Elsevier

1.5 CONCLUSIONS

It is interesting to note that the works of Yajima and Verbeek – superb examples of the bottom-up approach – were realized long before the concept of nanoscience was formulated by physicists. The studies on ceramic materials, using a combination of organic chemistry, polymer chemistry and catalysis, to surmount the difficulties of shaping an inorganic ceramic material, also show the unity of modern chemistry.

In conclusion, it is clear that chemistry possesses the essential synthetic and identification tools to contribute fruitfully to the advancement of nanosciences.

REFERENCES

[1] Nanosciences: au coeur des molécules. *Pour la Science* **2001**, December.
[2] M.L. Guillemot, C. Saout, R. Tardivel, C. Arnaud, A. Robin, *J. Eur. Hydrol.* **2002**, *33*, 1.
[3] R. Guilard, B. Roux-Fouillet, G. Lagrange, M. Meyer, A.T. Bucaille, *PCT Int. Appl.* **2001**, WO 2001046202.
[4] F. Cuenot, M. Meyer, A. Bucaille, R. Guilard, *J. Mol. Liq.* **2005**, *118*, 89.
[5] Y. Kishi, *Org. Chem.* **1988**, *1*, 253.
[6] Y. Kishi, *Pure Appl. Chem.* **1989**, *61*, 313.
[7] T. Yildirim, P.M. Gehring, D.A. Neumann, P.E. Eaton, T. Emrick, *Carbon* **1998**, *36*, 809.
[8] F. Richter, H. Vahrenkamp, *Angew. Chem.* **1980**, *19*, 65.
[9] H. Matsumoto, *IXth Int. Symp. Organosilicon Chem.* **1990**, *I21*, 35.
[10] W. Siebert, *Angew. Chem.* **1985**, *97*, 924.
[11] W. Siebert, *Pure Appl. Chem.* **1988**, *60*, 1345.
[12] J.J. Capponi, C. Chaillout, A.W. Hewat, P. Lejay, M. Marezio, N. Nguyen, B. Raveau, J.L. Tholence, R. Tournier, *Europhys. Lett.* **1987**, *3* (12), 1301.
[13] B. Chevalier, B. Buffat, G. Demazeau, B. Lloret, J. Etourneau, M. Hervieu, C. Michel, B. Raveau, R. Tournier, *J. Phys.* **1987**, *48*, 1619.
[14] K. Ariga, A. Vinu, M. Miyahara, *Curr. Nanosci.* **2006**, *2*, 197.
[15] D. Jérome, A. Mazaud, M. Ribault, K. Bechgaard, *C. R. Acad. Sci.* **1980**, *290*, 27.
[16] D. Jérome, M. Ribault, K. Bechgaard, *New Scientist* **1980**, *87*, 1209.
[17] H.S. Nalwa (ed.) *Handbook of Organic Conductive Molecules and Polymers*, John Wiley & Sons, Ltd, Chichester, **1997**.
[18] R. West, Polysilanes, *The Chemistry of Organic Silicon Compounds*, S. Patai, Z. Rappoport, (eds), John Wiley & Sons, Ltd, Chichester, **1989**, *19*, 1207–1240.
[19] B. Chaudret, *Top. Organomet. Chem.* **2005**, *16*, 233.
[20] J.-P. Majoral, A.-M. Caminade, *Top. Curr. Chem.* **2003**, *223*, 111.
[21] L. Griffe, M. Poupot-Marsan, J.J. Fournie, R. Poupot, C.O. Turrin, J.-P. Majoral, A.-M. Caminade, French Patent 2873715, **2006**.
[22] W. Verbeek, G. Winter, German Patent 2.236.078, **1974**.
[23] S. Yajima, J. Hayashi, M. Omori, Japanese Appl. 7550.223, **1975**.
[24] S. Yajima, J. Hayashi, M. Omori, Japanese Appl. 7550.529, **1975**.
[25] K. Shiina, M. Kumada, *J. Org. Chem.* **1958**, *23*, 139.
[26] M. Birot, J.-P. Pillot, J. Dunoguès, *Chem. Rev.* **1995**, *95*, 1443.
[27] C.A. Aitken, J.F. Harrod, E. Samuel, *J. Organomet. Chem.* **1985**, *C11*, 279.
[28] J.F. Harrod, Y. Mu, E. Samuel, *Polyhedron* **1991**, *10*, 1239.
[29] R. Corriu, M. Enders, S. Huille, J. Moreau, *Chem. Mater.* **1994**, *6*, 15.

2

Nano-Objects

This chapter introduces nano-objects, which constitute the basic elements of nanosciences. It also underlines the essential role played by chemistry in the bottom-up approach.

2.1 INTRODUCTION

The top-down and bottom-up approaches – mentioned in Chapter 1 and illustrated in Figure 2.1 – make up the nanosciences.

The top-down approach corresponds essentially to the research of miniaturization processes in electronics. Clearly, chemistry is not really involved in this approach. In contrast, the bottom-up approach, which aims to prepare materials with well-defined (optical, magnetic, mechanical, chemical, etc.) properties from nanometric-sized starting materials (molecules, clusters, etc.), corresponds perfectly to the chemists' know-how. In this approach, chemistry is needed on the one hand for the synthesis of the nano-objects, and on the other hand for the assembling and organizing of these nano-objects into nanomaterials.

Nano-objects are the building blocks of the materials of the future. They are chemical products (molecule, metal complex, cluster, etc.) with two additional attributes. First, the nano-object is specially synthesized considering a specific (optical, magnetic, electrical, mechanical or even chemical for catalysis or selective separation purposes, etc.) property. Secondly, the nano-object must have chemical functions permitting its transformation into a nanomaterial, that is to say a material having the desired properties and specially shaped for a specific application.

Molecular Chemistry of Sol-Gel Derived Nanomaterials Robert Corriu and Nguyên Trong Anh

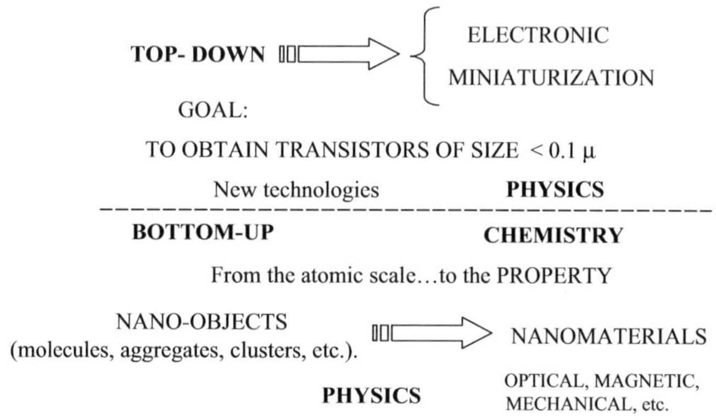

Figure 2.1 A schematic description of the two approaches of nanosciences. Reproduced by permission of *L'actualité chimique*

We shall return to this point in the examples to be discussed. Note that nano-objects are also used for fundamental studies of their properties in what is commonly called the 'exploration of the nanoworld'. This objective is very important for the potential discovery of unexpected properties. However, it concerns essentially physics. Only the synthesis of nano-objects is a matter for chemistry.[1–4]

2.2 PRESENTATION OF NANO-OBJECTS

In every sector of chemistry (organic, inorganic, macromolecular, etc.), syntheses have reached such an advanced state that chemists can now prepare any nano-object conceived for any properties.

A nano-object may be a simple molecule with a special property. Figure 2.2 shows two such examples corresponding to complexes of Cu^{2+} or Co^{2+} chelated by a tetraazamacrocycle (cyclam).[5] They have optical and paramagnetic properties. Chelation of a Eu^{3+} salt by phosphine oxides[6] leads to photoluminescent complexes. The diphenylphosphine used here has six phenyl groups, which play a double role. On the one hand, they induce an 'antenna effect': the luminous energy absorbed is transferred to the metal, thus amplifying the photoluminescence intensity. On the other hand, they provide a hydrophobic protection of the Eu^{3+} ion from complexation by H_2O molecules, which would inhibit the photoluminescence. In both cases, the $\Sigma[-Si(OEt)_3]$ groups allow the incorporation of these complexes into a silica matrix. Note that appropriate organic groups such as $-CH{=}CH_2$ would permit their integration in an organic polymer (polyethylene, polyacrylate, etc.).

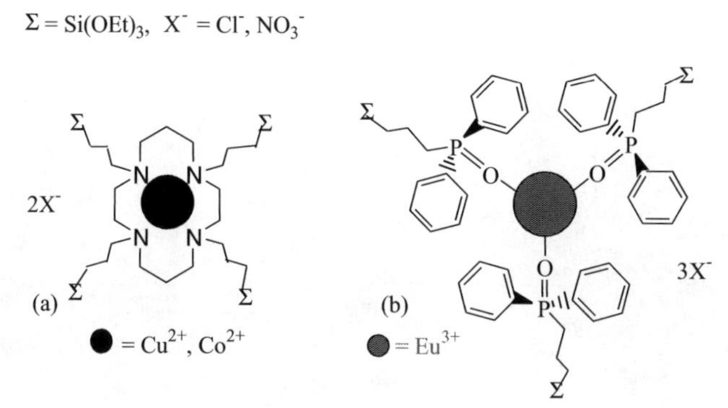

Figure 2.2 Two nano-objects having paramagnetic (a) and photoluminescent (b) properties

These two examples are taken from organometallic complexes, which constitute probably one of the richest 'mines' of nano-objects capable of leading to nanomaterials. There are several reasons for this. These complexes are nowadays numerous and diverse; synthesis of new chelatants is soaring and we know better and better the rules governing the complexation of metallic ions. Thanks to the progress of coordination chemistry, physical chemists are now able to study the physical properties of transition metal and lanthanide complexes. They have observed that these ions all have interesting physical properties, especially optical, magnetic and electrical properties. Thus the potential for creating new nano-objects and transforming them into nanomaterials is most promising.

Until now, however, the majority of nano-objects has been prepared for nanoworld exploration. The authors are not interested in their transformation into nanomaterials. Thus in the following examples, the nano-objects discussed do not possess the functional groups necessary for their transformation into nanomaterials.

Figure 2.3 shows a giant magnetic molecule with a 39/2 spin in the fundamental state. It corresponds to a $[Mn^{II}(Mn^{II}(MeOH)_3)_8][Mo^{V}(CN)_8]_6$ molecular cluster constructed with CN^- bridging ions.[7] This type of molecule, bound in an organized manner to a material,[8] can be used for storing magnetic information in the future.

Molecular electronics has been one of the driving forces of nanosciences. Studies in this field have significantly contributed to the recognition that nanosciences are one of the most fruitful interdisciplinary domains. One of the molecules used for studies of electronic transfer by scanning tunnel microscopy is intended for testing the current on

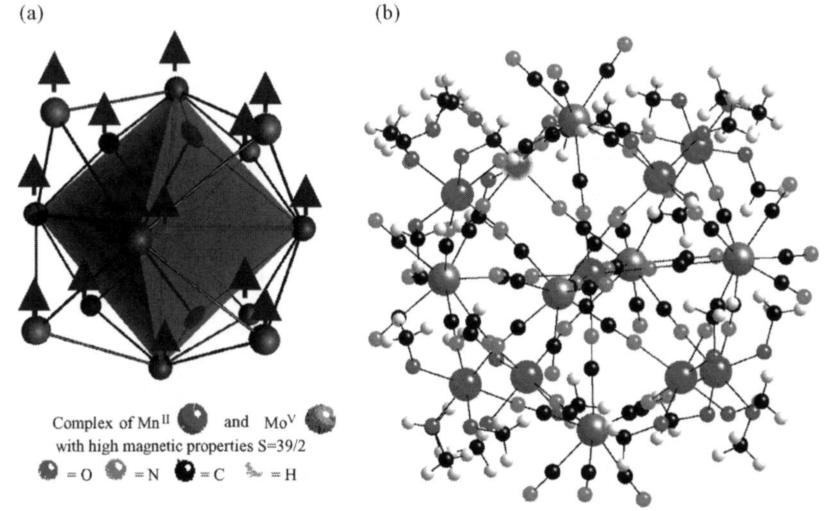

Figure 2.3 A bimetallic molecular complex having an important magnetic moment. The Mn^{II} and Mo^{V} ions are linked by $C{\equiv}N^-$ bridges. (a) Global scheme and (b) molecular structure. Reproduced by permission of *L'actualité chimique* (See Plate 2 for color representation)

a length of 4 nm (Figure 2.4). Its central portion, which is the conducting part, is essentially composed of aromatic rings (naphthyl) similar to those existing in graphite. The bulky *tert*-butyl groups attached at the extremities are the spacers, which isolate this molecule from the metal surface.[1]

Some nano-objects correspond to an even more complex approach. That is the case, for example, of molecular engines mimicking precise

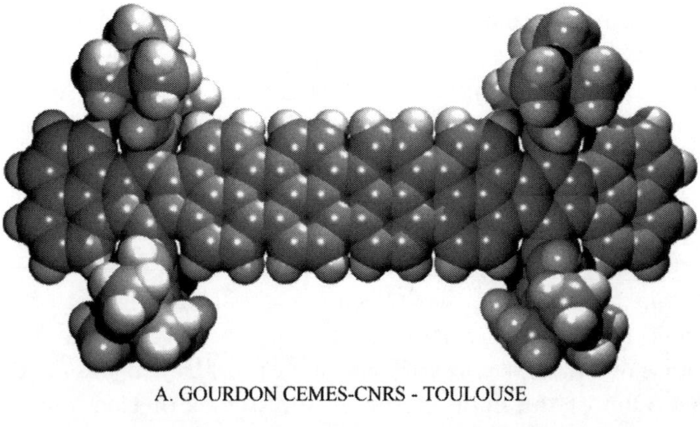

Figure 2.4 Nano-object permitting a single electron transfer by tunneling microscopy

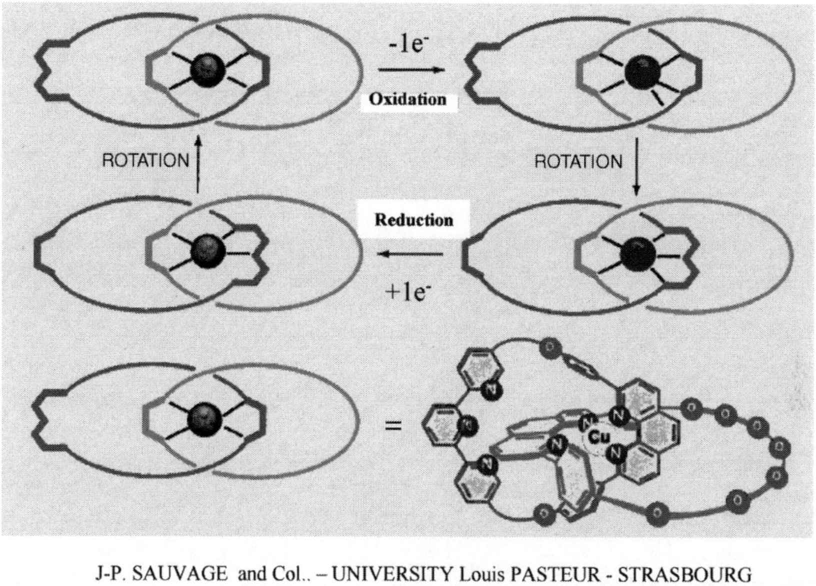

Figure 2.5 Molecular engine with a rotor capable of rotating inside a stator. Reproduced by permission of *L'actualité chimique* (See Plate 3 for color representation)

mechanical motions at the molecular scale. Figure 2.5 shows a rotating molecular engine.[9]

The motion is produced by the stability changes induced by oxidation (−1 electron) or reduction (+1 electron) in coordination complexes. In the reduced state (Cu^+) (2 + 2) tetracoordination is favored, whereas in the oxidized state (Cu^{2+}), one new coordination site is freed and (2 + 3) pentacoordination becomes preferred. The intramolecular motion enables the molecule to adopt the more stable conformation, going from (2 + 2) to (2 + 3) by oxidation or from (2 + 3) to (2 + 2) by reduction. It is the search for the energy minimum, which is responsible for the rotating motion.

2.3 SYNTHESIS OF NANO-OBJECTS

Synthesis of nano-objects may be very simple or quite complex. Thus the magnetic cluster (Figure 2.3) is obtained by simple crystallization of an appropriate mixture.[7] However, the molecule in Figure 2.5 requires a complex synthesis (shown in Figure 2.6). The strategy consists of preparing two links of the chain from common elements. They are then

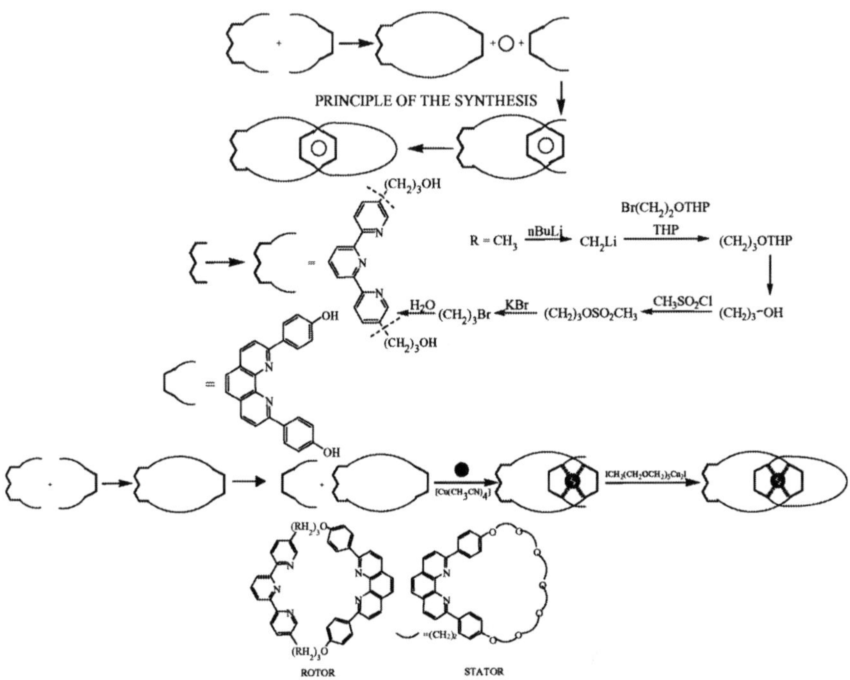

Figure 2.6 Synthesis of the molecular engine shown in Figure 2.5. Reprinted with permission from *Journal of the American Chemical Society*, Electrochemically Triggered Swinging of a [2] Catenate by Aude Livoreil, Christiane O. Dietrich-Buchecker, Jean-Pierre Sauvage,*116*, 20, 9399–9400. Copyright (1994) ACS

connected by complexation around a metal. In the last step, formation of two covalent bonds brings about the ring closure.

Synthesis of the molecule mimicking a molecular engine implies a total mastering of organic synthesis.[10]

2.4 THE NANO-OBJECT: ENTRY INTO NANOSCIENCES

In the bottom-up approach, the nano-object plays a double role. First, it serves in the exploration of the nanoworld. *Many nano-objects are, and will be, prepared in order to discover properties existing at the elementary scale,*[11–17] that is to say the scale of molecules, nanoparticles, clusters, etc. For example, the molecule shown in Figure 2.4 has been devised for the study of the longitudinal displacement of an electron by scanning tunnel microscopy.

Many of these properties will be different from those observed at the micrometric scale. Thus, in optics, the confinement phenomenon, [18] first discovered by Faraday,[19] and subsequently observed in metals and semiconductors can be mentioned. When the particles studied get too small (a few nanometers) some physical properties (in optics in particular) become different from the properties existing at the micrometric scale.[20–22] It is particularly obvious in the case of gold particles, whose color changes when the particle size becomes submicrometric.

The second function of nano-objects is to be the *starting product for nanomaterials*. These materials in the future must be contrived from nanometric entities, which have the desired property and can be assembled and shaped. Oversimplifying somewhat, we may say that the challenge in this domain is the *discovery of methods for organizing the matter in terms of properties*.

These properties can be physical (e.g. optical, magnetic, conducting, etc.), chemical (selective separation, catalysis, etc.), mechanical or even biological. Synthesis of nanomaterials thus requires two conditions: it is important that the nano-object be available by a proven chemical synthesis; and it must be obtained in sufficient quantities so that it can be transformed, shaped and studied as a nanomaterial. To illustrate, let us examine the examples mentioned before, namely the selective separation of lead in drinking water and the production of SiC ceramic fibers.

In both cases, the expected property is an intrinsic property of the nanomaterial: *specific chelation* of Pb^{2+} by 'tailor-made' cyclams in the first example (cf. Chapter 1) or high mechanical performance (SiC or Si_3N_4) *ceramics fibers* in the second example (cf. Chapter 1). Fibers were obtained by using a procedure, which allows a viscous crosslinked and spinnable polymer to be obtained. In both cases, the synthesis of the nano-object is only a preliminary stage, the objective remaining the nanomaterial resulting from a chemical assemblage of the nano-object, specifically functionalized for this purpose. In the first example, it is a simple grafting operation on a silica matrix. In the second example, synthesis and shaping require a series of specific chemical transformations.

Thus the nano-object, which is the basis of nanosciences, is much more than a simple chemical compound, as it must be *synthesized in terms of a specific property*. Such an objective may require extensive investigations.

2.4.1 Nano-Objects and the Exploration of the Nanoworld

When the nano-object is used for nanoworld exploration, the problem is the same. The nano-object synthesis must give a molecule suitable for a thorough study of a physical property. This study has been made possible by the discovery of observation techniques at the subnanometric scale, such as atomic force microscopy and scanning tunnel microscopy. These nanoscopic techniques permit the study of an elementary act such as the spin transition of an electron: this one-electron transistor case is a textbook example of molecular electronics. They also permit the displacement of metal atoms one by one on an appropriate surface. (We have already mentioned the company's acronym being written by IBM scientists.). It is equally possible to measure the energy sufficient to disrupt some biological interactions at the elementary scale.

We will not discuss at any greater length this area of nanosciences, which is essentially physics. Besides, there are already numerous journals dedicated to this topic. The role of chemistry in this field is reduced to the synthesis of the nano-object, which usually does not require any special invention. *As a matter of fact, existing synthetic tools permit the synthesis of all nano-objects conceived for a potential property.*

As already mentioned, the exploration of the nanoworld can lead to a new form of physics at the nanometric scale. It can also be very useful to chemistry and biology, permitting for example great advances in catalysis by offering precious information on the nature and the geometry of the elementary catalytic process.[4,23–33]

REFERENCES

[1] C. Joachim, J.-P. Launay, *Actualité Chimique* **2005**, *290–291*, 53.
[2] *Pour la Science* **2001**, December, 290.
[3] J.F. Gimzewski, C. Joachim, *Science* **1999**, *283*, 1683.
[4] *Nanocomposants et nanomachines*, Rapports 'Arago' de l'Observatoire Français des Techniques Avancées **2001**.
[5] G. Dubois, C. Reyé, R.J.P. Corriu, S. Brandès, F. Denat, R. Guilard, *Angew. Chem., Int. Ed.* **2001**, *40*, 1087.
[6] F. Embert, A. Mehdi, C. Reyé, R.J.P. Corriu, *Chem. Mater.* **2001**, *13*, 4542.
[7] J. Larionova, M. Gross, M. Pilkington, H. Andres, H. Stoeckli-Evans, H.U. Güdel, S. Decurtins, *Angew. Chem. Int. Ed.* **2000**, *39*, 1605.
[8] M. Verdaguer, A. Bleuzen, R. Lescouëzec, V. Marvaud, C. Train, *Actualité Chimique* **2005**, *290–291*, 44.
[9] J.-P. Collin, V. Heitz, J.-P. Sauvage, *Top. Curr. Chem.* **2005**, *262*, 29.
[10] J.-C. Chambron, S. Chardon-Noblat, C.O. Dietrich-Buchecker, V. Heitz, J.-P. Sauvage, *Macrocycle Synth.* **1996**, 207.
[11] C.P. Collier, G. Matterstei, W.E. Wong, Y. Luo, K. Beverly, J. Sampaio, F. Raymo, J.F. Stoddart, J.R. Heath, *Science* **2000**, *289*, 1172.
[12] A.R. Pease, J.O. Jeppesen, J.F. Stoddart, Y. Luo, C.P. Collier, J.R. Heath, *Acc. Chem. Res.* **2001**, *34*, 433.
[13] R.K. Soong, G.D. Bachand, H.P. Neves, A.G. Olkhovets, C.D. Montemagno, *Science* **2000**, *290*, 1555.
[14] J.-P. Sauvage, *Struct. Bonding (Berlin)* **2001**, 99.

[15] V.J. Langlais, R.R. Schlittler, H. Tang, A. Gourdon, C. Joachim, J.K. Gimzewski, *Phys. Rev. Lett.* **1999**, *83*, 2809.
[16] G. Dujardin, R.E. Walkup, P. Avouris, *Science* **1992**, *255*, 1232.
[17] B.C. Stipe, M.A. Rezaei, W. Ho, *Science* **1998**, *279*, 1907.
[18] M.A. El-Sayed, Acc. Chem. Res. **2001**, 257; A.P. Alivisatos, *Science* **1996**, *271*, 933; J. Shi, S. Gider, D. Babcock, D.D. Awschalom, *Science* **1996**, *271*, 937.
[19] M. Faraday, *Philos. Trans. R. Soc.* London **1857**, *147*, 145.
[20] M. Brust, M. Walker, D. Bethell, *J. Chem. Soc. Chem. Commun.* **1994**, 801.
[21] R.G. Nuzzo, D.L. Allara, *J. Am. Chem. Soc.* **1983**, *105*, 4481.
[22] C.B. Murray, D.J. Norris, M.G. Bawendi, *J. Am. Chem. Soc.* **1993**, *115*, 8706.
[23] *Electronique moléculaire*, Rapports 'Arago' de l'Observatoire Français des Techniques Avancées **1988**.
[24] *Nanotechnologies et micromachines*, Rapports 'Arago' de l'Observatoire Fran ais des Techniques Avancées **1992**.
[25] D. Eigler, E.K. Schweizer, *Nature* **1990**, *344*, 524.
[26] Y. Benenson, B. Gil, U. Ben-Dor, R. Adar, E. Shapiro, *Nature* **2004**, *429*, 423.
[27] C. Joachim, J.K. Gimzewski, A. Aviram, *Nature* **2000**, *408*, 541.
[28] F. Rosei, M. Schunack, P. Jiang, A. Gourdon, E. Legsgaard, I. Stensgaard, C. Joachim, F. Besenbacher, *Science* **2002**, *296*, 328.
[29] S. Fraysse, C. Coudray, J.-P. Launay, *Eur. J. Inorg. Chem.* **2000**, *7*, 1581.
[30] C. Joachim, J.K. Gimzewski, *Chem. Phys. Lett.* **1997**, *265*, 353.
[31] J.K. Gimzewski, C. Joachim, R.R. Schlittler, V. Langlias, H. Tang, I. Johannsen, *Science* **1998**, *281*, 531.
[32] F. Moresco, L. Gross, M. Alemani, A. Gourdon, C. Joachim, K.H. Rieder, *Appl. Phys.* A **2005**, *80*, 913.
[33] A. Carrella, J. Jaud, G. Rapenne, J.-P. Launay, *Chem. Commun.* **2003**, 2434.

3

Introduction to Material Chemistry

In the preceding introductory chapters, we set the stage for nanosciences and touched upon nanomaterials, their major objective. First and foremost, a nanomaterial is a material, which distinguishes itself only by its conception and its preparation at the nanometric scale. It is therefore important to survey now the vast domain of material sciences.[1–4] The purpose of this chapter is not to exhaustively review this subject, but only to point out its main characteristics and to highlight the concepts necessary to approach the synthesis and study of nanomaterials.

3.1 GENERAL REMARKS

Let us first emphasize that *nanosciences are going to revolutionize the habits and the ways that chemists think*. For a long time, chemistry has been divided into separate subdomains, *so that molecular chemistry, solid state chemistry and polymer chemistry have developed almost independently*. This state of affairs has permitted remarkable advances in each of these domains. However, what is new is that nanosciences and nanomaterials involve concepts, data, logic and methods coming from all three of these domains, so that they become completely intermixed.

Thus, until the last few decades, the quasi-totality of materials was prepared under thermodynamic control and at high temperatures (1000 °C and above). Obviously, these methods are not suitable for the preparation of nanomaterials. Indeed, nano-objects often contain

Molecular Chemistry of Sol-Gel Derived Nanomaterials Robert Corriu and Nguyên Trong Anh

organic, organometallic or biological entities which cannot survive temperatures higher than 400 °C. New methods are required. To simplify, one can say that the synthesis of nano-objects belongs to molecular chemistry and their assembling into nanomaterials is covered by organic and inorganic polymerizations, which allow operations at temperatures lower than 250–300 °C.

Compartmentalization and excessive specialization then become obsolete. A solid state chemist interested in nanosciences must learn the basics of polymer chemistry, organic chemistry and the methods of preparation of materials under kinetic control. Conversely, the synthetic chemist responsible for the preparation of nano-objects must have knowledge far beyond just molecular chemistry. Indeed, *a nano-object is not an objective in itself; it is only the first stage of a process leading to a nanomaterial, the real objective of nanosciences.* The nano-object must, from the beginning, contain functionalities for its assembly. This implies that the molecular chemist must be acquainted with the synthesis of organic and organometallic polymers. In addition, the chemist's expertise must include processing methods of solids as well as methods for their structural characterization. The chemist must also take interest in the solid texture (density, porosity, hydrophilicity, mechanical behavior, etc.).

Decompartmentalization is not limited to chemistry. The chemist cannot be indifferent to the (physical, mechanical, etc.) properties which direct the use of the nanomaterial. The chemist must interact with other scientists (physicists, mechanical engineers, etc.) capable of studying and exploiting these properties.

All this will be clearer as different types of materials are discussed. For now, let us only say that nanomaterials, the principal objective of the bottom-up approach, are solids endowed with useful physical properties. For this reason, they belong to material science, with *the know-how and capability implied by the term 'material'*.

3.1.1 The Difference Between Materials and Chemical Compounds

First, it must be stressed that the practical use of materials requires the *solid form* for obvious reasons. How could we conveniently transport electricity with liquids or construct computers with gaseous semiconductors? A material is designed for its intended use. In addition to its chemical characteristics and its physical, mechanical, chemical, etc. properties, the material must also be suitable for a specific use in precisely determined conditions.

Let us present some examples of 'use', a notion not too familiar to molecular chemists. Obviously, a mechanical piece intended for aeronautics and a biochip to be grafted on an organ must be dealt with differently. Nanomaterials for the selective separation of lead in drinking water (cf. Section 4.5) come as easily replaceable devices in which the selective chelatant is chemically grafted on an inert and porous support allowing easy passage of water. Electronic components are solid elements in which the functional parts lie on the surface. They are modular, interchangeable and designed to be easily connected and exchanged with one another. Materials for optical applications must be transparent in the window of the wavelengths concerned. Thus silica matrices should be preferred to glass matrices for applications in the violet and ultraviolet range, etc.

To cut a long story short, a material, associated in essence to a practical application, entails a number of constraints, the most important of which is its *shaping*. During this process, the material does not undergo any chemical change, but its morphology is radically altered. *In the course of its production, the material must go through a stage where it can take the form desired for the intended application.* The three principal forms used are that of a *fiber*, of a *film* or *coating* and of a *matrix* capable of incorporating a large variety of elements.

3.1.2 Examples of Shaping and Use

Optical fibers are a good example of *fiber materials*. They represent an important application for the future since optical transmission can convey more information than electronic transmission and, in addition, is more rapid and more precise. Optical fibers are made of pure silica drawn out to kilometer lengths, in a perfectly homogenous and calibrated manner. Such a technical feat implies a total mastering of the drawing out, which can be achieved by controlled fusion of a silica block of proven optical characteristics. A description of the technologies employed is outside the scope of this book.

Film materials are a widely explored domain, since films are well suited for the studies of physical, particularly optical, properties. They can also be used for the protection or functionalization of surfaces. Technologies have been perfected to control the homogeneity and thickness of the coating, the most important being dip coating and spin coating.

Dip coating consists of depositing the material in a viscous state (sol) on a surface (glass plate, silicon wafer, etc.). The support is plunged in the sol, and then removed at controlled speed, under controlled atmosphere (humidity, organic solvent, inert gas, etc.). The deposit thus obtained is even and has a thickness of 0.2–1 μm, which can be adjusted according to the viscosity of the sol and the speed used.

Spin coating can give thicker layers (0.5–5 μm). The sol is deposited on a glass plate which is submitted to rapid rotation. Centrifugal force enables a homogenous repartition and a rapid drying of the sample. Several successive layers can be deposited on the same sample.

Matrices permit the introduction into materials and the shaping of entities which cannot be directly used as materials because their morphology is not adapted to the intended use. This application will be exemplified by composite nanomaterials in Chapter 4.

3.2 INORGANIC MATERIALS: CRYSTALS AND GLASSES

Materials science began in the 1950s.[1–4] Scientists have mostly focused their attention on the preparation of inorganic solids and the study of their physical properties. Later, it was the useful properties which guided the search for new solids. This development has in fact given rise to *materials science* and transformed a large part of inorganic chemistry into the *physical chemistry of materials*.

Most of the studies in materials science concerned the solids obtained, under thermodynamic control, from a mixture of suitably chosen powders submitted to an appropriate thermal treatment. The physical properties associated with the chemical structures of these materials have been detected, and then investigated experimentally and theoretically. Thus, we can consider that, with the exception of organic polymers, a large majority of materials studied up to now (crystals, ceramics and glasses) were derived from the thermodynamic approach.

Ceramics are assemblages of microcrystals. Many physical properties require the existence of crystallinity extending over at least several micrometers. This is true for the photovoltaic property of TiO_2, with anatase as the crystalline phase. This is also the case for the magnetic property of iron oxides which depends strongly on the electronic environment of the iron atom in different crystalline states: hematite (α-FeO_3), maghemite (β-FeO_3), magnetite (Fe_3O_4). Finally, silicon is a semiconductor only on the surface of crystalline silicon.

Glasses are in a metastable amorphous state (there is a stable crystalline state corresponding to the same formula). Thus SiO_2 glasses are not birefringent, unlike SiO_2 crystals. Note, however, that the 'metastability' of glasses corresponds to a much longer life than that of 'stable' molecular compounds. Recall, for example, that most glasses from antiquity are still in a vitreous state. Thus glasses are markedly less prone to development or reorganization than kinetically controlled polymers.

Discussion of the chemistry of inorganic materials is outside the scope of this book for two reasons. First, there is an abundance of books and journals[1–4] adequately covering the subject. Secondly, the majority of the methods employed cannot yield nanomaterials. As a matter of fact, many of them require temperatures higher than 400 °C, which is totally incompatible with the presence of organic entities.

3.3 THERMODYNAMICALLY CONTROLLED ORGANIC-INORGANIC HYBRID MATERIALS

In addition to materials obtained by the thermal methods just mentioned, there exist hybrid nanomaterials which are inorganic materials containing organic components. These thermodynamically controlled crystalline materials can be divided into two groups: crystalline molecular materials and materials derived from hydrothermal synthesis. They shall be treated separately for they are prepared by different methods and do not possess the same characteristics.

3.3.1 Crystalline Molecular Materials

The best known example is that of crystalline stacks of tetrathiafulvalene (TTF) and of tetracyanoquinodimethane (TCNQ), often called Bechgaard salts.[5] These crystals have very high conductivity and their selenium analogs are even supraconductors.[6] Mention can also be made of the material obtained from TTF and a nickel complex $Ni(dmit)_2$, which exists as thin layers of assemblies of one unit of TTF for two units of Ni $(dmit)_2$.[7,8] The monocrystal conductivity at room temperature is 300 S cm^{-1}. This $TTF[Ni(dmit)_2]_2$ compound is one of the first metallic films derived from molecular materials (Figure 3.1). Under pressure, it becomes supraconductor.[9,10]

There are many materials obtained by thermodynamically controlled methods, whose structures are microcrystalline and, for this reason, show interesting physical properties. These properties are indeed closely connected with crystalline organization. However, in most cases, research

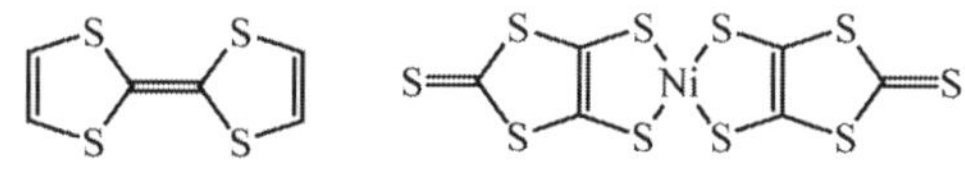

TTF Ni(dmit)$_2$

Figure 3.1 Structure of TTF[Ni(dmit)$_2$]$_2$ and photograph of its deposit on silicon. Reprinted with permission from *Advanced Materials*, Metallic thin films of TTF[Ni (dmit)$_2$]$_2$ by electrodeposition on (001)-oriented silicon substrates by De Caro, Dominique; Fraxedas, Jordi; Faulmann, Christophe; Malfant, Isabelle; Milon, Julie; Lamere, Jean-Francois; Colliere, Vincent; Valade, Lydie, *16(9–10)*, 835–838. Copyright (2004) Wiley-VCH

and development of the physical properties of materials[11–13] require samples of sizes much larger than a micrometer.[14]

These abundantly described materials[15–17] are outside the scope of this book.

3.3.2 Materials Derived from Hydrothermal Synthesis

Hydrothermal synthesis is a very general method for the preparation of crystalline solids, including those with quite complex structures. *Above all, it permits a reasonably predictive synthesis of thermodynamically controlled hybrid materials.*

Zeolith chemistry is a good example. Zeoliths are mixed oxides (for example SiO_2/Al_2O_3) frequently used in heterogeneous catalysis.[18–24] The methodology used for their synthesis has become widespread, for its versatility allows the assemblage of a wide variety of chemical entities. It permits the preparation of crystalline materials combining organic entities with suitably chosen inorganic units.[25] In practice, hydrothermal synthesis is conducted in an autoclave. The various components to be assembled are mixed with a diluent (generally H_2O). An acidic or basic catalyst is required. Additives which are not involved directly in the composition of the material are introduced in order to control the sizes of cavities and the crystalline type of the final product. This mixture is heated at temperatures from 100 °C up to several hundred degrees.

The mechanisms of hydrothermal synthesis are now better understood on account of NMR studies performed inside the autoclave. In particular, understanding the role of the organic entities used as 'templates' has improved their utilization. Thus, the most recent developments concerned predictive methods by computer, permitting a judicious choice of the building blocks for making the desired nanomaterials.[26–28]

As an example, the structure of a macroporous crystalline hybrid obtained by predictive hydrothermal synthesis is shown in Figure 3.2.

Finally, it is important to stress that this method allows the assemblage of inorganic entities presenting interesting physical properties. Thus the replacement of Si, Ge and Al by Fe, V and Ti gives permanent magnetization. Hydrothermal synthesis is, therefore, capable of coupling different

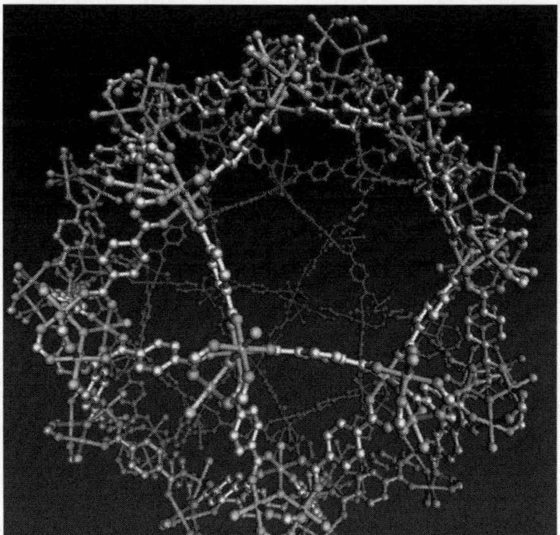

Figure 3.2 Structure of a hybrid material obtained by predictive hydrothermal synthesis (diameter 34 Å, volume 20 600 $Å^3$, hexagonal window 16.5 Å). Reprinted with permission from *Chemical Society Reviews*, Hybrid porous solids: past, present, future by Gerard Ferey, *37*, 1, 191–214. Copyright (2008) RSC (See Plate 4 for color representation)

properties, and thus belongs to the domain of interactive materials (cf. Chapter 6). Moreover, it is possible to adjust the material's properties by considering: (1) the inorganic framework; (2) the composition of organic spacers controlling the framework geometry and the sizes of pores; and (3) the nature of the entities introduced in the pores. [29–31]

These materials represent, so to speak, the thermodynamically controlled alternative to mesoporous materials which will be discussed in Chapter 6. In addition, they can incorporate organic molecules, macromolecules and even biological units.

3.4 CERAMIC MATERIALS OBTAINED FROM ORGANOMETALLIC POLYMERS: CERAMICS WITH INTERPENETRATING NETWORKS

A very special type of ceramics has been obtained by a molecular method analogous to the processes used for obtaining SiC and Si_3N_4[32] by Yajima[33,34] and Verbeek.[35] This method takes advantage of the well-known aptitude of bisacetylenes, $-C{\equiv}C{-}C{\equiv}C-$ (C_4), to react together to give a polycondensed network of enynes. Indeed, covalent bonds are made between sp carbon atoms belonging to two parallel chains, resulting in a sp^2 and sp carbon matrix which has been used for the ceramization of SiR_2 entities. Figure 3.3 shows the synthesis of polymers and the reticulation scheme induced by the autocondensation at 200 °C of bisacetylenic units (C_4).

Above 400 °C, elimination of CH_4 and H_2 and quantitative formation of a SiC/4C matrix in which SiC is embedded in a graphite type of carbon are observed.[36,37] Ceramization of amorphous SiC takes place between 500 °C and 600 °C with production of $^{\bullet}CH_3$ radicals, which abstract $H^{\bullet}$ from $Si{-}CH_2{-}H$, generating $Si{-}^{\bullet}CH_2$ and CH_4. The radical $Si{-}^{\bullet}CH_2$ then attacks $Si{-}CH_3$ with elimination of $^{\bullet}CH_3$ and the formation of $-Si{-}CH_2{-}Si-$ bonds in the carbon matrix (Figure 3.4). The two SiC and 4C phases remain intimately intermingled.* Crystallization of SiC accompanied by separation from the carbon matrix occurs only at 1400 °C, which shows that the mixing is particularly thorough. Another proof of this strong imbrication is the necessity of raising the temperature up to

* The Material is for this reason termed SiC/4C.

† At 600°C, Si_3N_4 is formed.

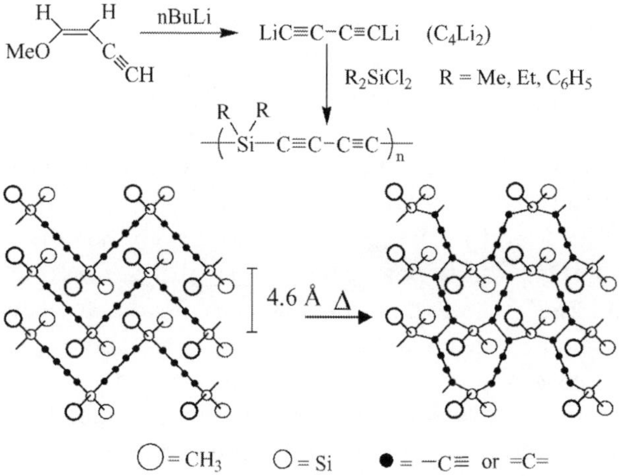

Figure 3.3 Schematic diagram of the synthesis of organosilicon polymers containing bisacetylenic C_4 units and their crosslinking at 200 °C with formation of enyne units by 1,4-addition. Reprinted with permission from *Angew. Chem., Int. Ed.*, Ceramics and nanostructures from molecular precursors by Robert J.P. Corriu, *39*, 1376–1398. Copyright (2000) Wiley-VCH

800 °C† under NH_3 atmosphere to achieve the elimination of carbon atoms as HCN.[38]

It has been possible to exploit the reducing power of the carbon matrix (4C) and its strong imbrication with SiC to obtain nanocomposite ceramics in which SiC is intimately associated with metal (Ti, Zr, V, Hf, etc.) carbides or nitrides. The method has been chosen to give a proper mixture at the nanometric scale.[39,40] As simple mixing of the SiC/4C powder with the oxides to be reduced was unsatisfactory, the following protocol has been set up. The oxide to be reduced (for example TiO_2) is mixed as a finely ground powder (1–5 μm) with a THF solution of the $[(CH_3)_2Si{-}C{\equiv}C{-}C{\equiv}C{-}]_n$ polymer. After elimination of the

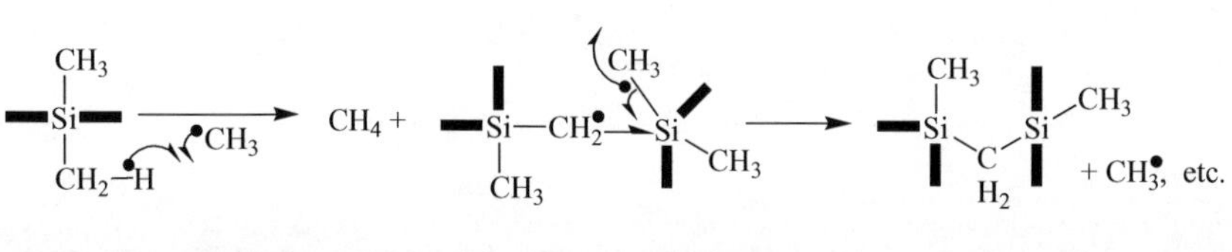

Figure 3.4 Mechanism of the initial steps of the ceramization process. Transformation of $>Si(CH_3)_2$ into SiC. Reprinted with permission from *Angew. Chem., Int. Ed.*, Ceramics and nanostructures from molecular precursors by Robert J.P. Corriu, *39*, 1376–1398. Copyright (2000) Wiley-VCH

solvent, the residue is pyrolyzed under argon. The temperature is raised progressively to allow the ceramization of SiC and the carboreduction of the TiO_2 oxide. Thermogravimetric analysis shows elimination of CH_4 and H_2 at 450 °C, corresponding to the ceramization of SiC. The system remains unchanged up to 1200 °C, where the carboreduction of TiO_2 occurs with elimination of CO. Interestingly, the carboreduction occurs between 1200 °C and 1400 °C for any oxide. Even the refractory oxides (ZrO_2 and HfO_2), whose powders are normally carboreduced between 2000 °C and 2500 °C, are transformed into carbides at 1400 °C. This is due to the intimate mixing of the oxide particle and the 4C phase: indeed, electronic microscopy shows that the oxide particles are encapsulated by the enyne units. This intimate blending of the supramicrometric oxide particles into the 4C network cannot be achieved by mixing the powders. The formation of SiC (400 °C) always precedes the carboreduction (between 1200 °C and 1400 °C) of the incorporated oxides (TiO_2, ZrO_2, HfO_2, V_2O_5, Nb_2O_5, B_2O_3) to give SiC/(TiC, ZrC, HfC, VC, etc.) composite ceramics.[41]

Nitrides can also be obtained by reaction of N_2 at high temperatures (~1000 °C) (Figure 3.5). This process occurs after the formation of SiC by reaction of N_2 on the oxides encapsulated in the 4C matrix. The SiC/2 (MN) (M = Ti, Zr, Hf) SiC/2/3Si_3N_4 mixed ceramics have been obtained in this manner. Note that nitrogen reacts directly on the metal when oxygen is eliminated by the sp^2 carbon matrix. This is a kinetic effect since gaseous N_2 reacts more rapidly than solid carbon with the metal. It is well known that transformation of carbides into nitrides by N_2 is impossible to achieve at any temperature.

It is interesting to note that these mixed ceramics constitute a category of materials never obtained before by thermal treatment of powders. The spectra show clearly the total absence of SiO_2 or other oxides. Moreover, the existence of carbides and nitrides is proven by X-ray diffraction. *In each case, the X-ray diagram corresponds to the sum of the diffraction*

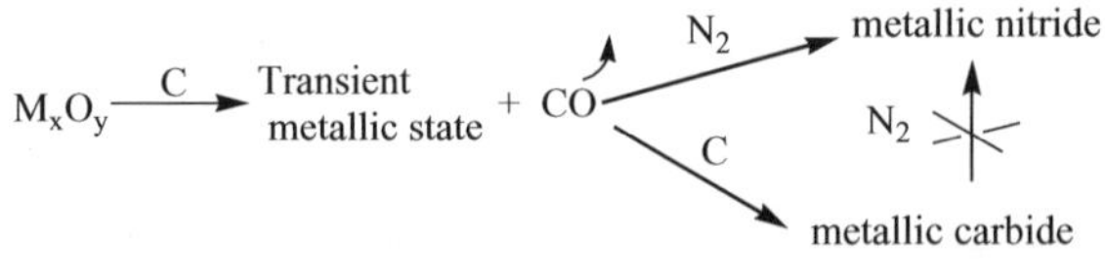

Figure 3.5 Schematic diagram of nitride formation by reaction of gaseous N_2; this is much more rapid than carbide formation by reaction with carbon. Reprinted with permission from *Angew. Chem., Int. Ed.*, Ceramics and nanostructures from molecular precursors by Robert J.P. Corriu, 39, 1376–1398. Copyright (2000) Wiley-VCH

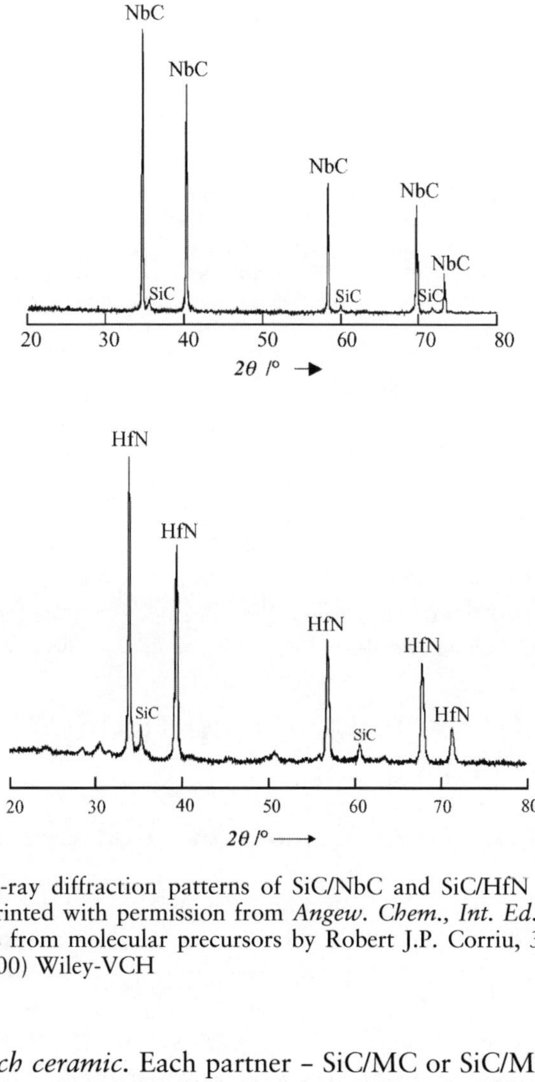

Figure 3.6 X-ray diffraction patterns of SiC/NbC and SiC/HfN nanocomposite ceramics. Reprinted with permission from *Angew. Chem., Int. Ed.*, Ceramics and nanostructures from molecular precursors by Robert J.P. Corriu, *39*, 1376–1398. Copyright (2000) Wiley-VCH

peaks for each ceramic. Each partner – SiC/MC or SiC/MN – is clearly identified, thus excluding the possibility of solid solutions or alloys (Figure 3.6). EDAX‡ diagrams show that the Si/M ratio corresponds to that in the reaction.

By analogy with polymers, these mixed ceramics can be described as ceramics with interpenetrating networks, in other words, they are intimate mixtures of two ceramics. They have been called *nanocomposite*

‡Energy dispersive analysis of X-Ray.

Table 3.1 Mixed SiC-MN nanocomposite ceramics

Polymer (equiv.)	Metallic oxide (equiv.)	Ceramic yield (%)[a]	Stoichiometry of mixed ceramic[b]
(1)	TiO_2 (2)	64 (62)	SiC-2 TiN
(1)	ZrO_2 (2)	66 (71)	SiC-2 ZrN
(1)	HfO_2 (2)	84 (81)	SiC-2 HfN
(5)	V_2O_5 (4)	58 (57)	5 SiC-8 VN
(5)	Nb_2O_5 (4)	71 (66)	5 SiC-2 $Nb_4N_{3.92}$
(3)	B_2O_3 (4)	49 (53)	3 SiC-8 BN
(3)	Al_2O_3 (4)	62 (62)	3 SiC-8 AlN
(1)	SiO_2 (2)	62 (59)	SiC-2/3 Si_3N_4

[a]Theoretical value calculated from oxide.
[b]Value obtained by IR and XRD.

ceramics. This work nicely illustrates the contributions of molecular chemistry in the ceramics field, by permitting access to materials impossible to obtain by other methods. Results obtained with the SiC-MN mixed ceramics are shown in Table 3.1.

3.5 INORGANIC POLYMER MATERIALS (SOL-GEL PROCESS)

3.5.1 Inorganic Polymerization: An Introduction

Inorganic polymerization, better known as the 'sol-gel process', is a general method for preparing oxides by the 'wet route' (i.e. in solution) at room temperature. As we shall see, it is more akin to molecular chemistry than to the chemistry of thermodynamically controlled materials. The sol-gel process was discovered in 1846 by Ebelmann.[42,43] It has benefited from developments inorganic polymerization.

3.5.1.1 Presentation of the sol-gel process

Inorganic polymerization, neglected for a long time, has experienced an extraordinary revival during the last two decades, as it corresponds to a *general method of preparation of oxides under mild conditions*. Let us recall that the usual methods of solid state chemistry require temperatures higher than 400 °C. *The importance of the sol-gel process is due to the fact that it allows the incorporation of* ***thermolabile*** *entities coming from all* *other fields* (organic, macromolecular, coordination, bioorganic and even biological) *of chemistry*. Let us try to summarize this long story.

Our account will be limited to the SiO_2 case,[44,45] which is the most frequently used material, being very versatile, stable and inexpensive. Reaction with water, at room temperature and in presence of a (acidic, basic or nucleophilic) catalyst, of a $Si(OR)_4$ (R = Me, Et, iPr, etc.) solution gives a transparent solid occupying the entire volume. This solid is not a precipitate, but a gel enclosing the solvent and sticking to the walls of the recipient. Equation (3.1) shows the global result of the hydrolysis of ethyl silicate and Figure 3.7 shows the various steps of the process.

$$Si(OEt)_4 + 2\,H_2O \xrightarrow{\text{Cat.}} SiO_2 + 4EtOH$$

Cat. = acid, base, nucleophile

Equation 3.1 Hydrolysis reaction of ethyl silicate

After a lapse of time which varies with the experimental conditions, a phase demixing occurs in which the solid SiO_2 expels the solvent and becomes denser. This stage is called *syneresis*. The solid obtained is ground, washed and vacuum dried to give a xerogel. Other treatments are possible. For example, hypercritical drying yields aerogels, which have valuable (dielectrical, insulating, etc.) properties. In spite of the interesting nature of these treatments, we shall discuss only the xerogels, which are the most used in the field of nanomaterials.

3.5.1.2 General characteristics of the sol-gel process

In practice, the sol-gel process is (at the macroscopic scale) very simple. *In one step and at room temperature, it transforms a molecule into a material*

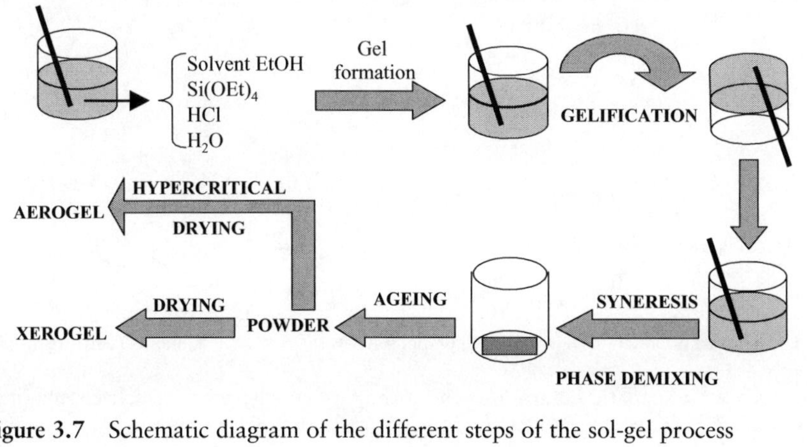

Figure 3.7 Schematic diagram of the different steps of the sol-gel process

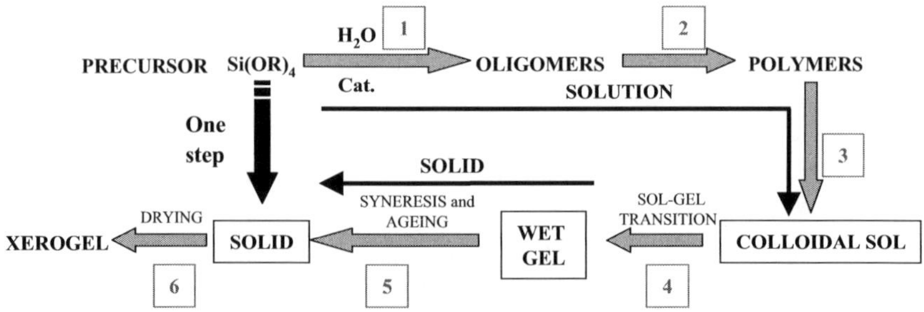

Figure 3.8 Schematic diagram of the chemical transformations occurring during the main steps of the sol-gel process

ready for shaping. At the nanoscopic and microscopic scales, it is in fact a very complex process consisting of several transformations of very different natures involving three states of matter: solution, colloid and solid.

Three stages can be distinguished. The first comprises steps 1, 2, and 3 (Figure 3.8) occurring in solution and corresponding to the formation of the Si–O–Si bonds. At first the precursor gives rise to nanometric linear and cyclic oligomers which, by polycondensations, gradually become intermingled to yield crosslinked polymers (10–100 nm). These polymers aggregate into micrometer-sized colloids, which are detectable by light diffusion. This colloidal solution is a viscous sol and corresponds to a crucial phase of the process (Figure 3.9). *It is the viscous sol which permits the shaping of the material.* It can be drawn into fibers, molded as a bulk solid or used as a covering by dip or spin coating.

In the second stage, the sol is solidified into a gel. The so-called sol-gel transition (step 4, Figure 3.8) step corresponds to a chemical reaction

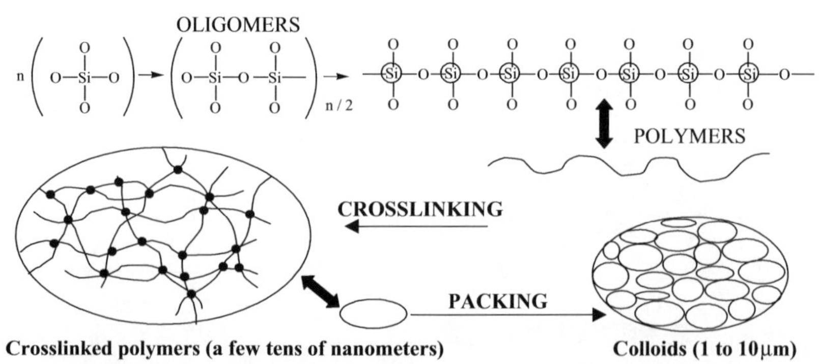

Figure 3.9 Schematic diagram of the formation of oligomers which change into crosslinked polymers, then into micrometric colloids constituting the sol

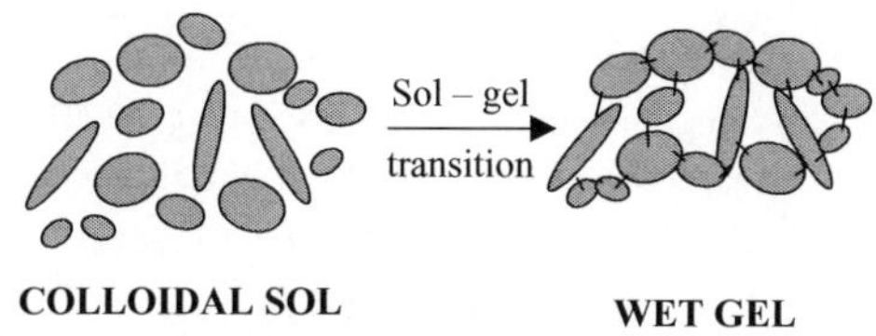

Figure 3.10 The sol–gel transition

occurring at the surface of the colloids (Figure 3.10). The rapid gelification can be easily understood: a few chemical bonds between the voluminous colloids, whose surfaces are covered with SiOH (or SiOR) groups, is enough to create a solid wide-meshed network capable of retaining all the solvent and of occupying all the space.

Step 5 (Figure 3.8) corresponds to a solid phase development. The number of bonds between the colloids increases, thus speeding up the densification of the solid. The initial stage, called *syneresis*, corresponds to the expulsion of the solvent. Continuation of this bonding process in the solid phase leads to an amorphous solid having stabilized granulometry and porosity, which depend on the experimental conditions. This ageing process is called the *Oswald ageing process*. It is very important because it controls the macroscopic characteristics of the solid (*i.e.* its *texture*, Section 3.5.2). Finally, step 6 corresponds to the drying of the washed solid. We shall consider the xerogel thus obtained as the final material. Note that the drying conditions can affect the material's texture significantly.

Let us examine now the molecular mechanisms governing the gelification process. Again, we shall discuss only the SiO_2 case, which is the most studied and the best known.

3.5.1.3 Study of the mechanisms governing the preparation of SiO_2 by the liquid phase sol-gel process

Inorganic polymerization is a process quite similar to organic polymerization. Its course is, however, much more complex, since polymerization can occur in four directions (corresponding to the four valences). The rates of these four polymerizations are relatively diverse and until now unknown. Organic and inorganic polymerizations both make use of molecular precursors and both are under kinetic control. Their results, therefore, depend appreciably on the experimental conditions.

Consider the main steps of the process allowing the preparation of SiO_2 by hydrolysis of a molecular precursor. The most general case is the hydrolysis of a tetra-alkoxysilane, $Si(OR)_4$ (R = Me, Et, iPr). We shall only discuss tetraethoxysilane, $Si(OEt)_4$ (TEOS), which is the most often used.

The two elementary reactions giving rise to the SiOSi bond are a hydrolysis reaction followed by a condensation reaction, both requiring a catalyst. Three types of catalyst are used: acidic (H^+), basic (OH^-) or nucleophilic [F^-, DMF (dimethylformamide), DMAP (*para*-dimethylaminopyridine), etc.].

The first steps are relatively well known [kinetic studies having been performed on simpler molecules, such as $RSi(OR)_3$ and $R_2Si(OR)_2$] and can, therefore, be discussed in detail[Equation (3.2)].

$$\equiv Si{-}OR + H_2O \longrightarrow \equiv Si{-}OH + ROH \quad \text{Hydrolysis}$$
$$\equiv Si{-}OH + \equiv Si{-}OH \longrightarrow \equiv Si{-}O{-}Si\equiv + H_2O \quad \text{Homocondensation}$$
$$\equiv Si{-}OH + \equiv Si{-}OR \longrightarrow \equiv Si{-}O{-}Si\equiv + ROH \quad \text{Heterocondensation}$$

Equation 3.2 Polycondensation reactions

When acid-catalyzed, the first reaction occurs on the alkoxysilane conjugated acid [Equation (3.3)].

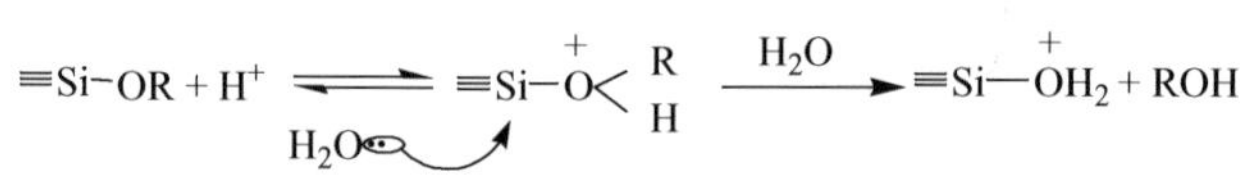

Equation 3.3 Mechanism of the acid hydrolysis of an alkoxysilane

The second reaction is that of the Si-OH oxygen either on the conjugated acid of the alkoxysilane (heterocondensation) [(1), Equation (3.4)] or on the conjugated acid of the silanol (homocondensation) [(2), Equation (3.4)].

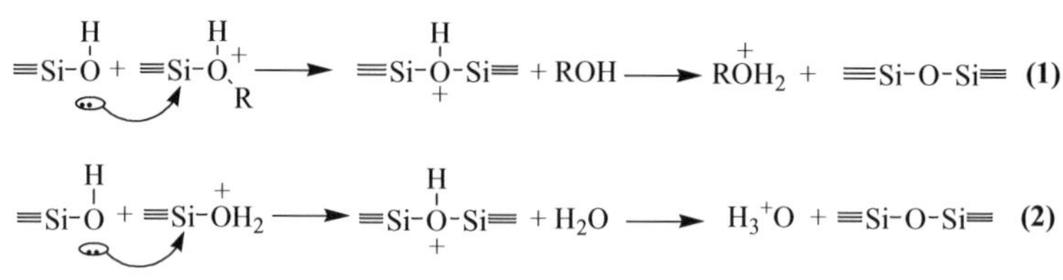

Equation 3.4 Mechanism of the acid polycondensation. (**1**) Heterocondensation. (**2**) Homocondensation

Basic catalysis has been depicted as a nucleophilic attack of the OH^- ion on the Si atom with displacement of RO^-. The silanol thus obtained, being more acidic than an alcohol, is deprotonated by RO^-, producing a silanolate, which in turn can attack either an alkoxysilane molecule (heterocondensation) or a silanol molecule (homocondensation) to give the Si–O–Si bond [Equation (3.5)].

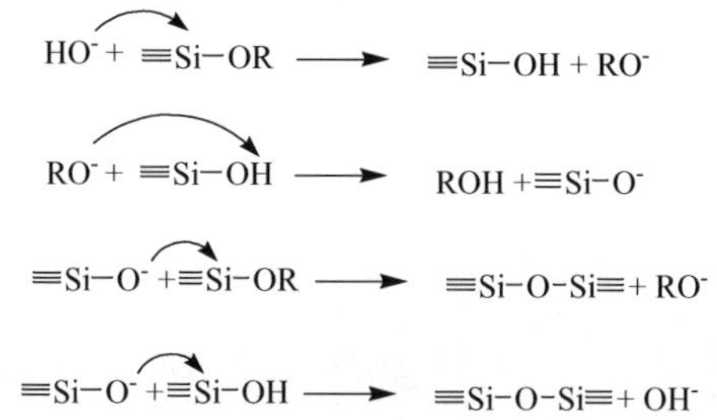

Equation 3.5 Mechanism of the basic polycondensation

Nucleophilic catalysis operates by extension of coordination on the silicon atom, followed by nucleophilic attack of H_2O on the pentacoordinated complex (Figure 3.11).

We have just described the initial steps of the process transforming Si $(OR)_4$ into $(RO)_3Si{-}O{-}Si(OR)_3$. The system becomes rapidly very complex because: (1) each silicon atom can react in four directions; and (2) hydrolysis competes with the polycondensation reaction. At the present time, even an approximate description of this process is impossible. At best, we can propose a reasonable scheme showing in a qualitative manner the different reactions which may be involved in the process. The key steps are the hydrolysis and the polycondensation on each functional group.

Figure 3.12 shows a schematic diagram of the polycondensation process. The first steps give rise to the reactive intermediates A, B, C. These reactions are accompanied by redistribution reactions between Si-OH and Si-OR, as shown by the equations A ⇆ B and B ⇆ C. Cyclizations simultaneously occur. All these reactions are equilibrated. They are very dependent on the experimental conditions, especially on the nature and concentration of the catalyst. For instance, redistribution equilibria are favored by nucleophilic catalysis much more than by acid catalysis.

Thus, from the initial formation of the oligomers, the polycondensation can be directed by the experimental conditions toward the formation of chains, cycles or aggregates. Our present knowledge does not allow us to

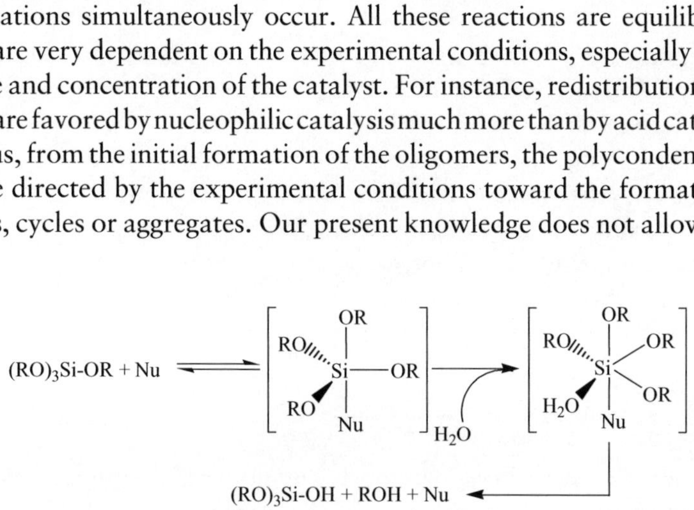

Figure 3.11 Nucleophile-activated hydrolysis mechanism of a tetra-alkoxysilane

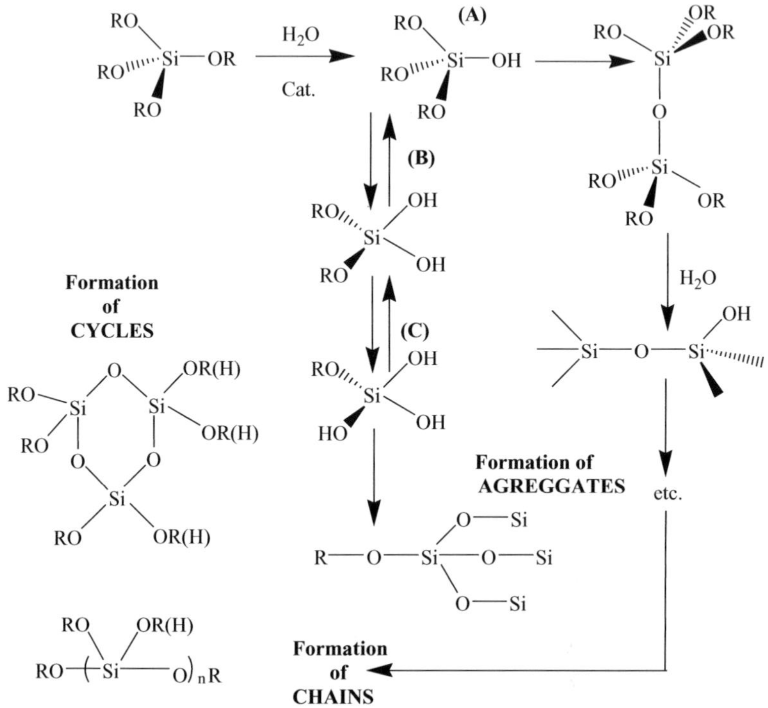

Figure 3.12 Schematic diagram of the reactions which may occur in the first steps of a tetra-alkoxysilane hydrolytic polycondensation: hydrolysis, condensation, redistribution, and cyclization. Reprinted with permission from *Angew. Chem., Int. Ed.*, Ceramics and nanostructures from molecular precursors by Robert J.P. Corriu, 39, 1376–1398. Copyright (2000) Wiley-VCH

discriminate between the different routes. Many studies have been done, but no general rule has been formulated. The only noteworthy work is that published by Chojnowki and colleagues.[46] Until now, it is the only reliable evaluation of the relative rates of acid hydrolysis of the four Si–OEt functions of ethyl silicate, $Si(OEt)_4$. Unfortunately, these measurements are only valid for the experimental conditions (concentrations of the reactants, temperature, solvent, etc.) used by the group. Polycondensation remains an empirical domain: we know how to orient the reaction but do not completely understand it.

3.5.1.4 Evolution in the solid phase

The Oswald ageing process previously mentioned is even more difficult to control than the complex kinetics of the polycondensation reactions. It corresponds to *changes in the solid phase (Figure 3.8, step 5) in which the surface and the volume of the solid undergo important modifications*. The solid gel, formed in step 4 (sol–gel transition), occupies the entire volume and contains all the liquid. It slowly rearranges by expelling the solvent (syneresis). This modification may be slow but it is important and irreversible. The texture of the solid, as measured by parameters such as specific area, porosity, density and hydrophilicity, etc., essentially depends on it. After drying (step 6), a xerogel of amorphous silica is obtained, whose centesimal composition provides little information. The only available data are the previously mentioned textural characteristics and that from ^{29}Si NMR, which can provide an estimate of the relative degree of substitution on the silicon atom: Q^0, **Si**$(OR)_4$; Q^1, (SiO)**Si**$(OR)_3$; Q^2, $(SiO)_2$**Si**$(OR)_2$; Q^3, $(SiO)_3$**Si**(OR); Q^4, $(SiO)_4$**Si** (see Section 3.5.4).

Note finally that the catalyst affects the ageing process. This phenomenon is far from understood. However, some observations have been made, which form the rudiments of the kinetic control we shall discuss later. For instance, it is known that a nucleophilic catalyst (for example F^-) favors the redistribution reactions on the silicon atom and that these reactions speed up the ageing of the material. In contrast, basic catalysis induces a partial equilibrated dissolution of the silica. This is a well known phenomenon: at $pH > 8$, silica can be dissolved in water according to the equilibrium:

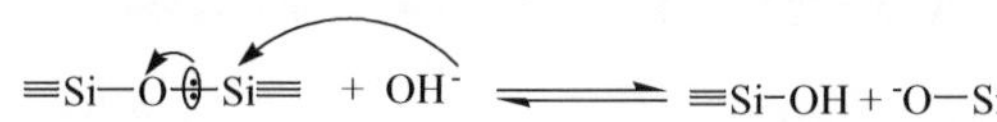

Equation 3.6 Basic cleavage of SiOSi bonds

3.5.1.5 Versatility of inorganic polymerization

In the preceding sections, we discussed the hydrolytic polycondensation of alkoxysilanes. Bonds other than Si-OR[47–53] can be used, for example $Si{-}NH_2$ or Si–NH–R which can be hydrolyzed in acid media (Figure 3.13). However, Si–Cl leads to precipitation.

It may be convenient sometimes to employ Si–H bonds, since they can be hydrolyzed using basic or nucleophilic catalysts (NH_4F and nBu_4NF being the most efficient) to give polysiloxanes. F^- ions react by nucleophilic activation extending the coordination of the silicon atom:[54] the SiH bond of a pentacoordinated complex can be substituted by the H_2O nucleophile. It is also possible to use activation by transition metals, which corresponds to a mechanism in which the metal catalyst (Pd here) is inserted in the SiH bond. This leads to a lengthening of the SiH bond, thus facilitating the attack by H_2O (Figure 3.14).

$$\equiv Si{-}NHR + H_2O \xrightarrow{H^+} \equiv Si{-}OH + RNH_2$$

two-step process

$$\equiv Si{-}NHR + H^+ \rightleftharpoons \equiv Si{-}\overset{+}{N}HR(H)$$

$$\equiv Si{-}\overset{+}{N}HR(H) + H_2O \longrightarrow \equiv Si{-}OH + RNH_2 + H^+$$

Figure 3.13 Simplified representation of the chemical reactions occurring in the inorganic polymerization of an aminosilane by acid hydrolysis

Other hydrolyzable functions exist, the best known being Si–O–COR, Si–SH and Si–SR. They are less convenient to use, but may prove interesting for the preparation of mixed materials in which silica is combined with other oxides or matrices.

This versatility is very important as the silicon polycondensation can then be adapted to many different situations.

3.5.2 Physical Characteristics of the Solid Obtained

3.5.2.1 Physical parameters characterizing the texture

The solid obtained by the sol-gel process, once dried, is silica in various forms, generally a powder. The so-called *texture* of the material can be

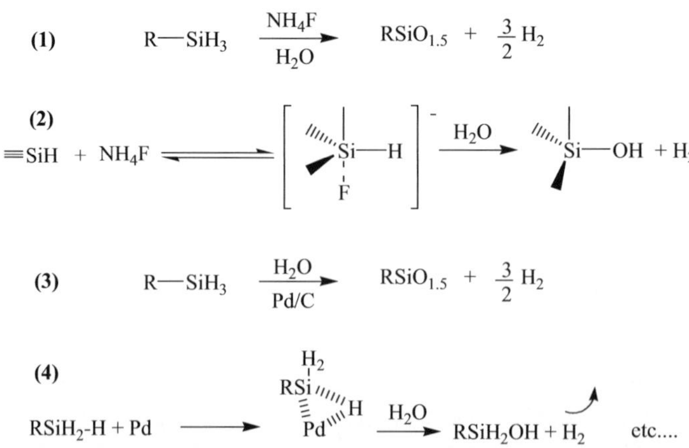

Figure 3.14 Mechanistic scheme for the Si-H bond hydrolysis activated by F^- [(1), (2)] and by Pd [(3), (4)]. Structure of the reactive intermediates

described and characterized by several macroscopic quantities. The most usual parameters are: the *density*; the *granulometry* (which gauges the size and size dispersion of the particles); the *specific surface area* [expressed in $m^2 g^{-1}$, which measures the total (internal and external) surface]; the *porosity* (which quantifies the total porous volume, particularly the distribution and the sizes of pores); and the *hydrophilicity* (which determines the capacity of the material to absorb water vapor).

The texture may be quite different, depending on the materials. Thus, the specific surface area of a silica powder may vary over a large range, from a nonsignificant surface ($<10\,m^2 g^{-1}$) for resinous materials, to values exceeding $1000\,m^2 g^{-1}$. The hydrophilicity ratio in silicas may vary from 1 to 5 depending on the experimental conditions chosen. For the porosity, the pore sizes may fluctuate from 1 nm to several micrometers (Figure 3.15). It may be centered on one or two values (monodisperse) or present a large distribution of sizes (polydisperse). The granulometry is generally far from uniform. The density of the material also varies, since it depends directly on its granulometry and its porosity.

Clearly, the textural characteristics strongly depend on the experimental conditions. The method (reagent concentrations, nature and concentration of the catalyst, solvent, temperature, etc.) is very important since every parameter is involved, for one reason or another, in the two stages of the process: the reactions in solution and the gel ageing in the solid. Note that the drying process also influences the texture. In fact, every experimental factor, including those apparently insignificant, can play a nonnegligible role. This is often the cause of disagreement between authors, as the experiments are difficult to reproduce because of the overly succinct description of the method. *Only a precise experimental procedure carried out with the utmost care, and the precaution of a sufficient ageing time can ensure the production of a stable and reproducible material.*

Due to the difficulties encountered in their texture control, the solids obtained by the sol-gel process have been erroneously described as

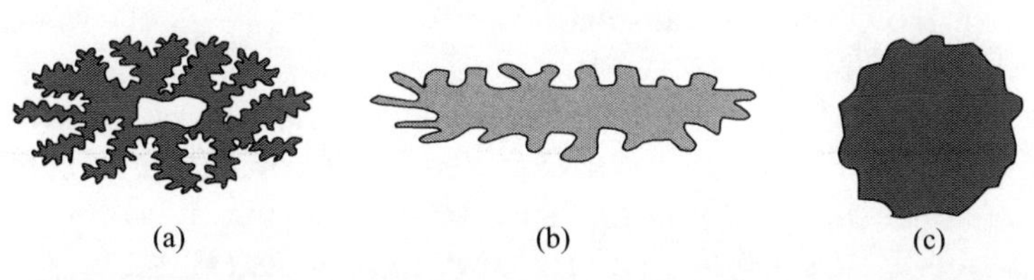

Figure 3.15 Schematic diagram of particles presenting different types of porosity: (a) high internal and external porosities; (b) normal surface porosity; (c) poor surface porosity

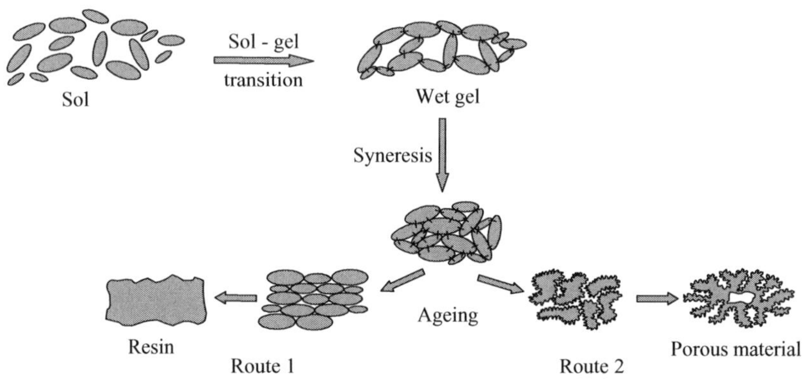

Figure 3.16 Outline of the formation of the specific surface area and the porosity during the ageing step: formation of a resin (route 1); formation of a porous solid (route 2). Reprinted with permission from *New J. Chem.*, Supramolecular self-organization in non-crystalline hybrid organic-inorganic nanomaterials induced by van der Waals interactions, Lerouge, Frédéric; Cerveau, Geneviève; Corriu, Robert J.P., *30*, 1364–1376. Copyright (2006) RSC

unstable. In fact, they are *materials under kinetic control*. They seem to evolve because modifications of their surface and volume occur during the Oswald ageing process and continue during the drying process. This is due to the polycondensation and redistribution reactions taking place within the Si–O–Si network, which affect the properties of the texture.

Figure 3.16 shows the different stages which may be envisaged for the formation of the solid from the colloidal solution. For the sol–gel transition to take place, a small amount of polycondensation is enough (e. g. formation of a single Si–O–Si bond between two colloidal particles of the sol). The gel corresponds to a large-mesh network in which the solvent is retained. Continuation of the polycondensation process reduces the size of the cavities and promotes the solvent expulsion (syneresis). Thus, during the ageing of the solid, the porosity and the specific area slowly change until a solid, stable under the ambient conditions, is obtained. However, several months are sometimes necessary to achieve this stabilization.

The modifications of the texture, due to the kinetic control, are incompatible with what can be expected from a material, which *must be stable to be practically useful*. This is the reason why it is important to submit the *aged solid to a hydrothermal synthesis* (100–130 °C in an autoclave for 24 h in the presence of an acid or nucleophilic catalyst). This operation stabilizes the material by completing the redistribution and polycondensation reactions.

3.5.2.2 Importance of the texture of solids

There exists a whole field of research on solid texture, particularly on the granulometry and the size of the pores. These precise data are fundamental for some scientific studies and also for numerous applications. For example, the fineness and properties of a ceramic will depend not only on the size of the particles but also on the homogeneous repartition of the sizes. A monodisperse repartition is mandatory for a quality ceramic.

In the same manner, the properties of a heterogeneous catalyst are strongly related to the interactions between the catalyst and the surface of the support (usually an oxide: SiO_2, TiO_2, ZrO_2, etc.). Figure 3.17 illustrates this point: the OH groups located at the surface of the oxide are ligands which tie up the metal but still allow free access to the reactants. This explains why the size, the geometry of the pores of the oxide support and the number of the OH groups as well as their distribution are fundamental for the efficiency of heterogeneous catalysis.

The inorganic charges in polymers are another example of the importance of solid texture. They are responsible for the mechanical properties of the material, and so they must be perfectly adapted to the intended application. Their efficiency depends essentially on the state of their surface, their density and their porosity. Car tires are a good example. Indeed, without carbon black charges, (natural) polyisoprene and (synthetic) polybutadiene rubbers cannot show the expected mechanical properties. It is the carbon black-polymer combination which enables the tires to resist the pressures and deformations experienced. A 'green' tire owes its longevity to the partial replacement of carbon black by silica particles, which confer to the tire a better resistance to abrasion. These silica particles must, beforehand, be chemically treated to mask their hydrophilicity, and to permit their incorporation into a hydrophobic (carbon black, polyisoprene and polybutadiene) material. Figure 3.18 shows the surface modification of the SiO_2 particles: the SH groups permit their insertion by forming

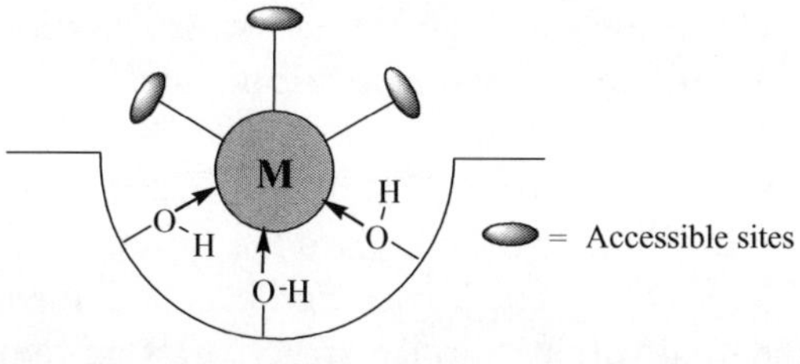

Figure 3.17 Schematic diagram of a catalytic site corresponding to a metal atom coordinated on a solid support

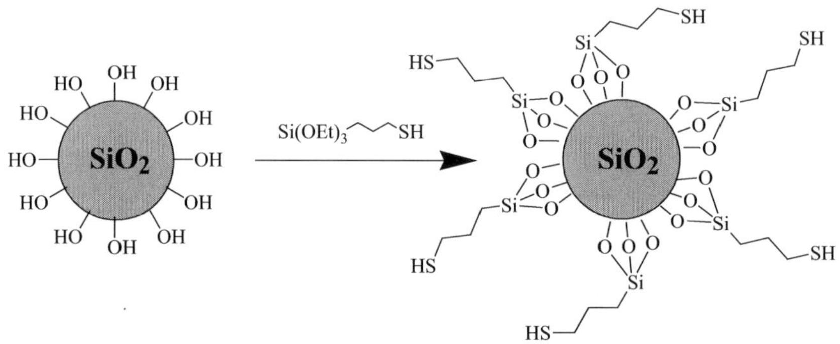

Figure 3.18 Schematic diagram of the surface modification of the silica incorporated into 'green' tires

covalent bonds during the vulcanization (*i.e.* the crosslinking of polyisoprenes and polybutadienes by sulfur atoms) of the polymers.

In conclusion, in materials sciences, it is crucial to take into account the macroscopic structural characteristics of the solid. These parameters determine, to a great extent, the properties used in their applications.

3.5.2.3 Determination of the texture of materials

As specialized books exist on this subject,[55] we shall discuss only briefly the methods used to determine the texture of materials.

For the purpose of this book, the most important parameters are those describing the material at the nanometric scale: the density, the hydrophilicity, the *total specific surface area*, the sizes of the pores and their distribution.

Densities in the liquid or the gaseous phases are measured using a pycnometer. Hydrophilicity is determined by weighing the water vapor adsorbed. For silica, it varies from 8 to 50%. The measurement is done in a desiccator under vacuum and in the presence of saturated water vapor controlled at 60% humidity (saturated NaBr solution). Obviously, hydrophilicity depends on the porosity but also on the number and distribution of the OH groups on the surface of the solid. Therefore, the chemistry of the formation of the solid will determine this quantity.

The so-called B E T[56] (from the initials of the authors Brunauer, Emmett, and Teller) and B J H[57] (Barrett, Joyner, and Halenda) methods permit the evaluation of the total specific surface area, expressed in $m^2\,g^{-1}$, and the estimation of the porosity from the absorption and desorption isotherms of a gas (usually N_2). We shall not go into detail about these measurements,

Table 3.2 Texture of some silica obtained under different experimental conditions[58]

Catalyst	S_{BET}[a] ($m^2 g^{-1}$)	S_{meso}[b] ($m^2 g^{-1}$)	V_P[c] ($cm^3 g^{-1}$)	V_μ[d] ($m^3 g^{-1}$)	$\bar{D}_P$[e] (Å)
NaOH	380	310	1.03	0.03	110
HCl	410	10	0.19	0.16	18
NH_3	350	180	0.56	0.07	63
NBu_4F	510	270	0.86	0.10	67

[a]BET specific surface area.
[b]BJH mesoporous surface area.
[c]Total porous volume.
[d]Microporous volume.
[e]Mean diameter of pores.

their accuracy and their precise meaning. Let us stress only the importance of these parameters on the properties of powders. As already mentioned, a material is more than a chemical compound, its *use* depends mostly on its intrinsic physical properties. Various applications (charges for polymers, supports for catalysts, solids for gas adsorption, protective coatings, etc.) are possible according to the size, the regularity of the pores and of the global specific surface area (Table 3.2).

Remember that the total specific surface area includes the external surface of the particles and the surface of the pores inside the material. Values can vary from 0 to 1500 $m^2 g^{-1}$. Low specific surfaces correspond to resins whose internal and external porosities are usually negligible. The sizes of the pores are also quite diverse, from the ultramicropores (a better name would be nanopores) which vary from 0.2 to 3 nm, to macropores of micrometric sizes (which should be called micropores). The importance of the texture of materials vindicates the next section, devoted to texture control.

3.5.3 Control of the Texture of Materials

3.5.3.1 General remarks

The previous pages may give the impression that the complexity of the inorganic polymerization process would represent an insurmountable obstacle for its use in the development of nanomaterials.

Let us recall that the preparation of nanomaterials (which contain thermolabile organic or biological components) is not compatible with thermal methods which use temperatures largely exceeding 400 °C. With the exception of hydrothermal synthesis, inorganic polymerization (sol-gel process) remains the only alternative capable of incorporating nano-objects into inorganic matrices. The materials thus obtained were *kinetically controlled* solids, which for this reason were considered as unstable materials. The apparent variability of the texture seemed hopeless: indeed, *a material whose development cannot be controlled is not reliable.*

However, the difference between an unstable product and a kinetically controlled product *is not a simple question of terminology*. We do not know how to deal with an unstable product. On the other hand, kinetic control means only that *the texture and the structure* (cf. Chapter 4) *of the final product (in this case the material) depend on the parameters governing the kinetics of its formation during the inorganic polymerization as well as during the ageing process*. Therefore, it should be possible to optimize them by varying these parameters, as we shall now demonstrate. Moreover, on account of the sol-gel process, it will be possible *to go from the nano-object to the shaped nanomaterial in one single step at room temperature*. It is, therefore, particularly important to study the conditions permitting the control of the texture of these materials, to make them reliable and usable at temperatures lower than 200–250 °C.

It suffices to take advantage of the experience acquired in molecular chemistry. The situation is, in fact, exactly that of a chemist trying to optimize a (kinetically controlled) organic, organometallic or polymerization reaction. In all these cases, the studied reaction is competing with many other unwanted, even antagonistic processes. To favor the desired reaction, it is necessary to adjust the experimental parameters. Therefore *the kinetic control of inorganic polymerization provides chemists with the means for optimizing the material's texture*. In other words, kinetic control has positive aspects which should be put to good use.

3.5.3.2 Texture and kinetic parameters: some examples

In this section, we shall examine hybrid nanomaterials, which are the only materials whose texture has been systematically studied as a function of the parameters governing the kinetics of the polycondensation on the silicon atom. Similar studies have never been done on silica.

These nanomaterials – which will be more extensively discussed in Chapter 5 – are obtained by polycondensations of $-Si(OR)_3$ groups linked to an organic entity (spacer). They constitute a single phase since the organic and inorganic components are joined by Si–C bonds. These

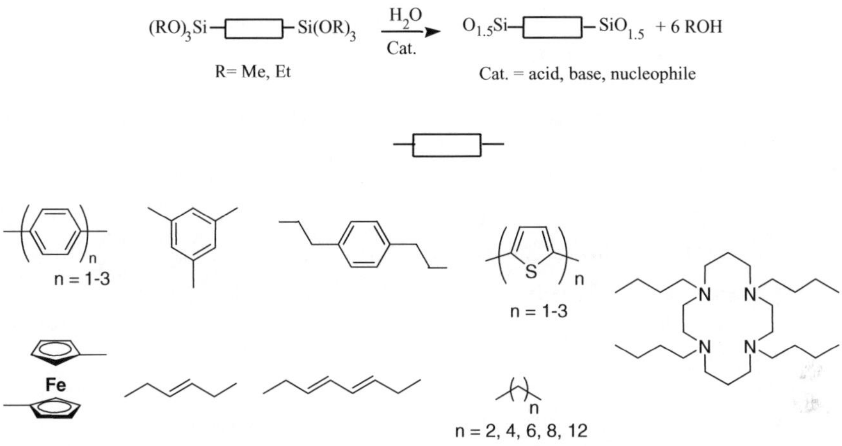

Figure 3.19 Polysilylated molecular precursors used to study the influence of kinetic parameters on the material's texture. Reprinted with permission from *New J. Chem.*, Supramolecular self-organization in non-crystalline hybrid organic-inorganic nanomaterials induced by van der Waals interactions, Lerouge,Frédéric; Cerveau, Geneviève; Corriu, Robert J.P., *30*, 1364–1376. Copyright (2006) RSC

covalent bonds resist hydrolysis and are preserved during the polycondensation process, as shown by ^{13}C and ^{29}Si NMR of the final product. For the moment, we shall present only some examples illustrating the possibility of controlling the texture of the solids (Figure 3.19).

It has been shown that the texture of the material obtained, as measured by specific surface area, porosity and adsorption–desorption isotherms, is a function of the kinetic parameters (temperature, nature and concentration of the catalyst, solvent, etc.). The nature of the spacer has also proved to be important.

These parameters are numerous and it is necessary to list them exhaustively at this stage:

- **Nature of the catalyst:** acid (HCl), base (NaOH), nucleophile (F^-, DMAP, DMF, etc.).[59–67]
- **Nature of the leaving group:** usually OR ($R = CH_3$, Et, iPr). However, other groups may be used: NHR or NH_2 which involves acid catalysis; or H which requires either a nucleophilic or a metal catalyst (Figure 3.14).[63,64,68,69] Cl is used very seldom.
- **Nature of the solvent:** usually the alcohol which is the conjugated acid of the leaving group (EtOH, MeOH). THF is also used. Other solvents, dioxane, formamide, DMF, etc.[59–62,65,66,70–72] can be employed.
- **Nature and concentration of the reagents:** H_2O, substrate, catalyst.[73,74]

- **Temperature:** this parameter has seldom been studied although it can play a very important role in the polycondensation phase as well as in the development of the solid phase.[64,66,75–78]
- **Geometry of the molecular entity:** rigid, flexible, linear, planar, tetrahedral, etc.[79–84]

One of the most illustrative examples is that of the solids obtained from two very similar precursors [(**1**) and (**2**), Figure 3.20]. Hydrolysis of these precursors carried out in a homogeneous phase, *in exactly the same conditions*, gives rise to two materials having *completely different* characteristics.[65]

Furthermore, as will be discussed in Chapter 5, these materials present very different chemical behaviors which can be explained by the existence of a structural organization in the solid. This organization is induced by van der Waals lipophilic interactions between the organic entities (*cf.* Chapter 5).

The specific surface area of (**2**), which comes generally as a nonporous resin, may be markedly increased by changing the experimental conditions. At room temperature and with TBAF as catalyst, a porous solid is obtained (25% of micropores smaller than 20 Å, the remaining pores ranging between 20 Å and 120 Å), with a specific surface area of 565 m^2 g^{-1}. All the other conditions remaining the same, simply by raising the temperature to 110 °C, a solid with a monodisperse porosity of 100 Å and a specific surface area of 780 $m^2\,g^{-1}$ is obtained.[66]

Likewise, the specific surface area of (**1**) can be adjusted by using the nature and concentration of the catalyst. For example, with a nucleophilic

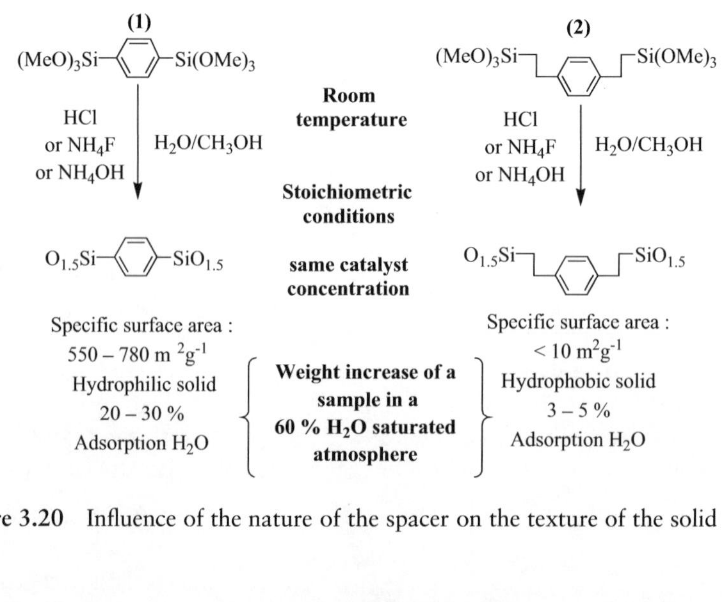

Figure 3.20 Influence of the nature of the spacer on the texture of the solid

Table 3.3 Effect of the nature of the solvent in the case of two precursors with different structures

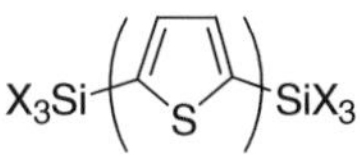

1

2

Precursor	X	Conc. (mol. l^{-1})	Solvent	Catalyst (Conc.)[a]	Temp. (°C)	BET Spec.S.($m^2\,g^{-1}$)
1	OMe	1	MeOH	NH_4F (1)	20	685
		1	THF	NH_4F (1)	20	20
2		0.4	EtOH	HCl (10.8)	20	<10
	OEt	0.4	THF	HCl (10.8)	20	380

[a]Concentration in molar %.

Reprinted with permission from *Chem. Mater.*, Nanostructured organic-inorganic hybrid materials: kinetic control of the texture, Cerveau, Geneviève; Corriu, Robert J.P.; Framery, Eric, *13*, 3373–3388. Copyright (2001) American Chemical Society

catalyst such as (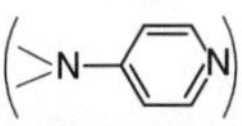) dimethylaminopyridine (DMAP), the specific surface area can vary from 0 to 875 $m^2\,g^{-1}$.

The hydrophilicities of (**1**) and (**2**) can also be modified: (**2**) is a hydrophobic material whose weight increase is always smaller than 4%. It is possible, however, to increase this value up to 8 or 9% by using DMAP as a catalyst and by polymerizing at temperatures higher than 30 °C.[66] In contrast, (**1**) comes as a solid with hydrophilicity ranging between 30% and 40%. This can be lowered to 9% by using nucleophilic catalysts at 110 °C in TBAF.[66]

To demonstrate how large the range of possibilities can be, the effect of a certain number of experimental parameters on the texture of materials obtained with different precursors is shown below. With a few exceptions, experiments were conducted at room temperature.

Table 3.3 illustrates the influence of the solvent on the specific surface area for two different structures (one rigid and one flexible), all other parameters being kept constant.[62,74]

Tables 3.4 and 3.5 show the importance of ageing on the texture formation (specific surface area, porosity, etc.) and on the level of polycondensation at silicon estimated by NMR.[76,77,85]

Table 3.6 establishes the influence of temperature,[§] during the inorganic polymerization step in the liquid phase, on the level of polycondensation at silicon, the specific surface area and the porosity.[66,75] These results were obtained with different precursors.

[§]The influence of this parameter has not been studied in the case of SiO_2.

Table 3.4 Effect of the ageing time

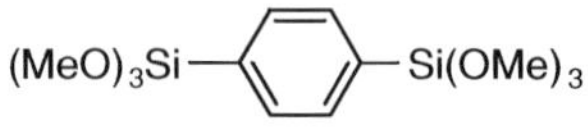

1

2

Precursor	Catalyst	Ageing (days)	^{29}Si RMN (%) Cond.[a]	Spec. S. ($m^2 g^{-1}$)	BET μ-pores (%)	Pores size (Å)
1	DMAP	0	60	<10	–	–
		3	60	475	65	20–60
		6	64	680	65	20–70
		15	67	875	45	20–50
2	TBAF	0	85	<10	–	–
		6	90	290	5	20–70
		15	90	350	5	20–70
		30	90	410	0	45

[a]Cond., Level of condensation at Si.

Table 3.5 Change in the level of condensation at silicon during ageing. This level is calculated according to the equation: cond. = [0.5(area T^1) + 1.0(area T^2) + 1.5 (area T^3)]/1.5(see Section 3.5.4)

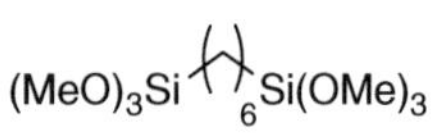

1

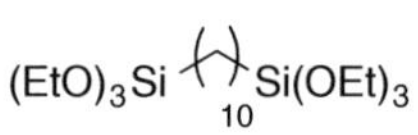

2

Precursor	Solvent	Cat.	Temp. (°C)	Ageing (days)	^{29}Si RMN(%)*									
					Gel					Xerogel				
					T^0	T^1	T^2	T^3	Cond.	T^0	T^1	T^2	T^3	Cond.[a]
1				2	0	17	64	19	67	0	20	55	25	68
	THF	HCl	20	4	0	18	64	18	67	0	18	64	18	67
				6	0	15	71	14	66	0	15	71	14	66
				30	0	9	67	24	72	0	8	67	25	72
				2	0	45	55	0	52	0	32	53	15	61
2	THF	HCl	20	4	0	35	55	10	58	0	26	51	23	66
				6	0	30	58	12	61	0	23	51	26	68
				30	0	6	69	25	73	0	3	63	34	77

[a]Cond., Level of condensation at Si.

Table 3.6 Effect of the polycondensation temperature on the texture in the case of precursors having different geometries

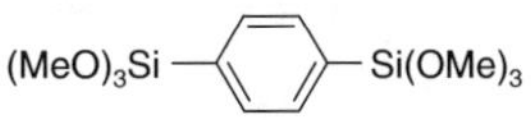

1

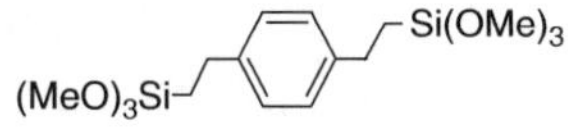

2

3

Precursor	Solvent	Catalyst	Temp. (°C)	^{29}Si RMN (%)[a] Cond.[b]	BET		
					Spec. S. ($m^2 g^{-1}$)	μ-pores (%)	Pore size (Å)
		TBAF	−20	71	900	55	20–120
1	THF		20	67	1300	10	20–120
			55	74	1300	5	55
			110	76	1230	5	45
		DMAP	−20	58	<10	–	–
			30	64	680	65	20–70
			55	62	580	55	20
2	THF	TBAF	−20	82	<10	–	–
			20	87	565	25	20–120
			55	80	700	5	40
			110	81	780	0	100
		HCl	20	73	<10	–	–
			55	73	<10	–	–
3	THF	TBAF	−20	81	250	20	20–50
			20	86	665	10	20–60
			55	91	870	5	35, 60
			110	93	920	0	35, 55, 80

[a]cf. Section 3.5.4
[b]Cond., Level of condensation at Si.
Reprinted with permission from *J. Mater. Chem.*, Sol-Gel process - influence of ageing on the textural properties of organosilsesquioxane materials, Cerveau, Geneviève; Corriu, Robert J.P.; Framery, Eric, *11*, 713–717. Copyright (2001) RSC

Table 3.7 presents the effect of temperature on the specific surface area and the porosity, during the ageing process.[85]

Polymerizations carried out at low temperature in the liquid or solid phase lead to resinous materials having low porosities and specific surface areas. In contrast, all the ageing conducted at higher temperatures gives porous systems whose porosity may be monodisperse (Figure 3.21).

Finally, Table 3.8 demonstrates the possibility of controlling the specific surface area by varying the polycondensation parameters.[60]

3.5.4 Solid State NMR: A Very Useful Tool

Solid state NMR is a totally reliable tool for the characterization of materials. Using the magic angle spinning and polarization transfer

Table 3.7 Effect of the ageing temperature in the solid phase in the case of precursors having different geometries

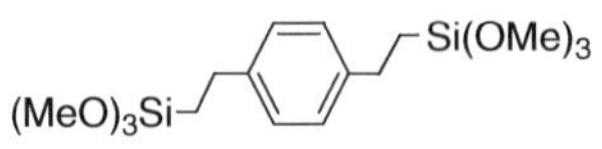

1

2

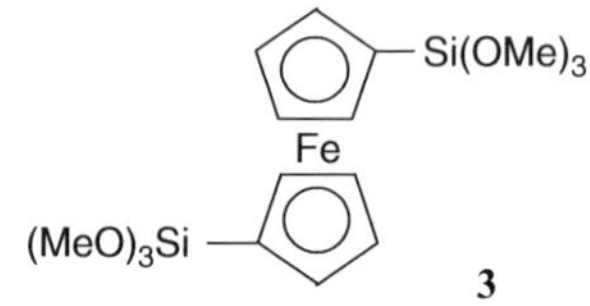

3

Precursor	Solvent	Temp. (°C)		BET		
		Gel	Ageing	Spec. S. ($m^2 g^{-1}$)	µ-pores (%)	Pore size (Å)
1	THF	−20	−20	<10	–	–
		−20	55	715	5	40
2	THF	−20	−20	<10	–	–
		−20	55	865	0	80
3	THF	−20	−20	<10	–	–
		−20	55	525	45	20–50
		55	−20	220	35	20–80
		55	55	275	40	20–120

techniques, one can obtain the NMR spectra for a great number of nuclei (^{13}C, ^{29}Si, ^{17}O, ^{11}B, ^{31}P, ^{15}N, ^{19}F, ^{119}Sn, etc.).

In our case, the two most important nuclei are ^{29}Si and ^{13}C. The first serves to characterize the level of polycondensation at silicon in silica and in silylated organic-inorganic hybrid materials. The second is necessary for the characterization of the organic part of these hybrids.

Among other nuclei, ^{31}P is one of the most reliable probes: not only is ^{31}P the major isotope but, in addition, it gives widely separated signals, which are, therefore, easily detectable.

We shall now outline the information that can be obtained from ^{29}Si NMR, which is the most frequently used for the nanomaterials we shall discuss.

The ^{29}Si nucleus, with a spin of ½, is not abundant (4.7%). Its relaxation time may also be very long, especially in solids (some materials have relaxation times in the order of 30 min). These two characteristics make ^{29}Si a nucleus difficult to exploit experimentally, and polarization transfer techniques are necessary (see below). One of the most important pieces of information is the level of polycondensation at silicon. ^{29}Si NMR is sensitive to the second neighbor and so permits, at least qualitatively, the detection of the different types of silicon atoms present in the material.[86–88] In the hydrolysis of a tetra-alkoxysilane, $Si(OR)_4$, five types of silicon differing in the degree of polycondensation can be distinguished.

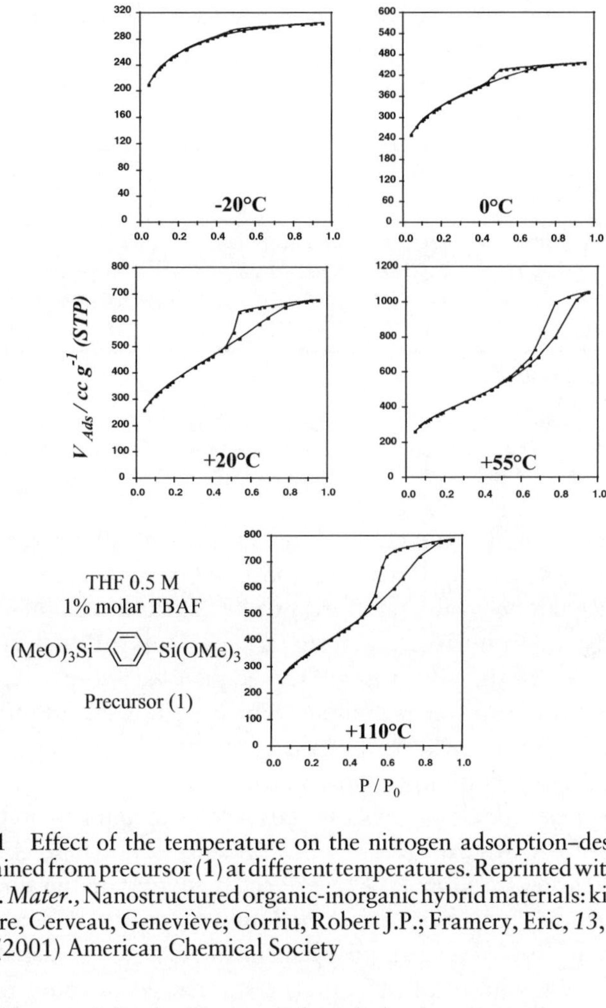

Figure 3.21 Effect of the temperature on the nitrogen adsorption–desorption isotherms obtained from precursor (**1**) at different temperatures. Reprinted with permission from *Chem. Mater.*, Nanostructured organic-inorganic hybrid materials: kinetic control of the texture, Cerveau, Geneviève; Corriu, Robert J.P.; Framery, Eric, *13*, 3373–3388. Copyright (2001) American Chemical Society

In hybrid materials, the Si atom is bonded at least to one atom of carbon and the four types of polycondensation (T^1, T^2, T^3, with T^0 corresponding to the precursor) are shown in Figure 3.22.

Similarly, in the case of substitution by two carbon atoms, the D^0, D^1 and D^2 units can be easily identified.

Thus the Q^0, Q^1, Q^2, Q^3, Q^4; T^0, T^1, T^2, T^3 and D^0, D^1, D^2 signals correspond to distinct ^{29}Si NMR chemical shifts. Furthermore, each of the Q^n, T^n and D^n groups can be distinguished from the others in the NMR spectrum.

In most cases, the presence of SiOH or Si–O–C does not significantly modify the position of the ^{29}Si NMR signal while a Si atom belonging to the Si–O–Si sequence results in a definite chemical shift. This allows an easy interpretation of the spectra since substitutions by OR and OH will

Table 3.8 Control of a material's texture by modification of experimental parameters

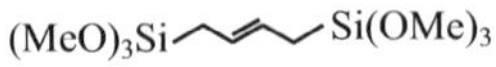

1

Precursor	Conc. ($mol\,l^{-1}$)	Solvent	Catalyst (conc.)[a]	Temp. (°C)	BET Spec. S. ($m^2\,g^{-1}$)
1	1	THF	NH_4F (0.1)	20	20
	1	MeOH[b]	NH_4F (0.1)	20	<10
	3[b]	MeOH	NH_4F (0.1)	20	325
	0.1[b]	MeOH	NH_4OH[b] (2.6)	20	980

[a]Concentration in molar %.
[b]Modified parameter.
Reprinted with permission from *Chem. Mater.*, Nanostructured organic-inorganic hybrid materials: kinetic control of the texture, Cerveau, Geneviève; Corriu, Robert J.P.; Framery, Eric, *13*, 3373–3388. Copyright (2001) American Chemical Society

not be differentiated. Only the polycondensation at silicon can be evaluated by type Q signals for silica and type T or D signal for the hybrids. All these signals can be easily identified on the NMR spectrum.

In contrast, quantitative estimations are much more difficult, due to the low abundance of the ^{29}Si isotope. Consequently, the spectra are taken with polarization transfer techniques which permit the amplification of the ^{29}Si signal. They are known by the acronym CPMAS (cross polarization magic angle spinning) (Figure 3.23).

With this method, chemical shifts can be obtained, but not a quantitative evaluation of the degree of polycondensation. These data can be considered as an estimate of the level of crosslinking on the silicon atom. However, studies have been performed in order to obtain a more precise evaluation. Spectra of a number of hybrid solids have been studied in the absence of polarization transfer. It appears that deconvolution methods permit a relatively accurate estimation of the level of polycondensation at silicon.[71,89,90]

To illustrate the contribution of NMR to the understanding of the development of the solids obtained by inorganic polymerization, we present one of the few works which show the progress of the polycondensation at silicon during the drying process of a SiO_2 gel. Figure 3.24 shows the modification on going from the wet gel to the dry gel (3 h, 120°C) and the influence of different catalysts.[58] Unfortunately these kinds of NMR experiments are highly time consuming.

In this case, the spectra realized without polarization transfer permit a quantitative estimate of ^{29}Si. A clear evolution of the level of polycondensation has been observed during the drying of the solid material. These

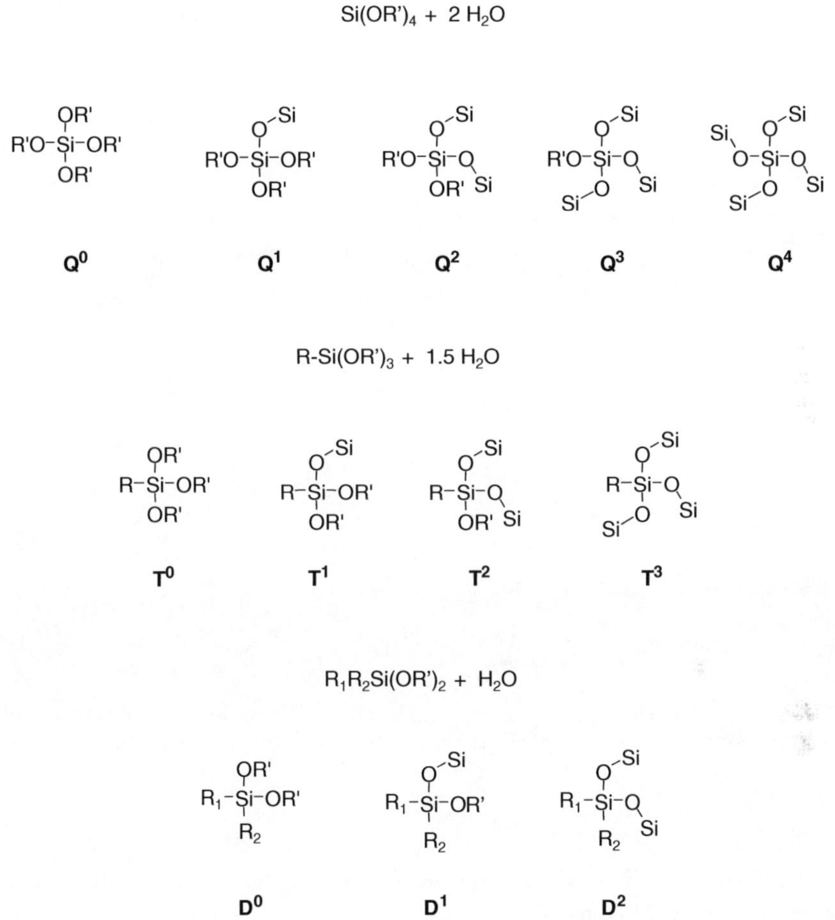

Figure 3.22 Schematic diagram of the different entities that may be formed in the polycondensation of $Si(OR')_4$ (Q), $RSi(OR')_3$ (T), and $R_1R_2Si(OR')_2$ (D). R′ = Me, Et, iPr, H

results show clearly that the polycondensation at silicon is not completed in the liquid phase. It continues during the ageing of the solid and during the drying process which involves a heat treatment.

3.6 INORGANIC POLYMERIZATION AND MOLECULAR CHEMISTRY

Inorganic polymerization (sol-gel) offers an alternative to inorganic synthesis using thermal routes. As we have seen, it is a kinetically

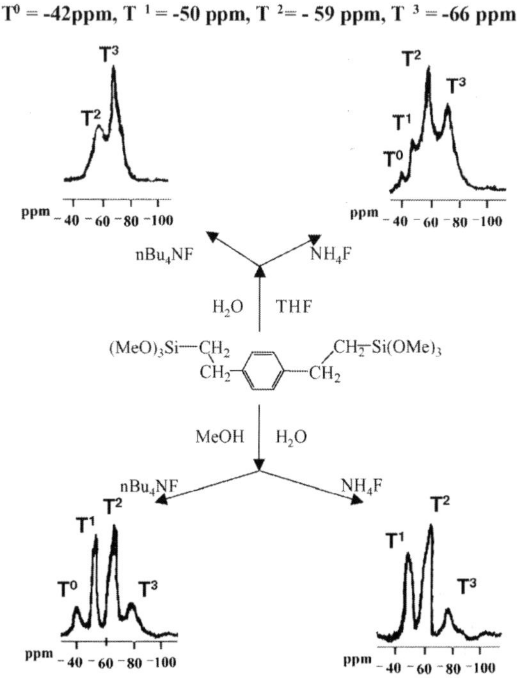

Figure 3.23 Influence of the catalyst on the level of condensation at silicon. Reprinted with permission from *The Chemistry of Organosilicon Compounds,* vol. 3, Z. Rappoport and Y. Apeloig, Boury, B.; Corriu, R.J.P., chap. 10, p. 565. Copyright (2001) John Wiley & Sons, Ltd

controlled process producing oxides by hydrolysis of molecular precursors in mild conditions.

Although silica is always taken as an example, it should be pointed out that this methodology opens up three promising prospects: the total compatibility between silica and molecular systems; the extension to other oxides; and the generalization to other types of inorganic combinations.

3.7 SILICA AND MOLECULAR CHEMISTRY: A DREAM TEAM

The development of chemistry in the bottom-up approach depends on *the transformation of nano-objects into nanomaterials.* Existing methods are adequate for nano-object synthesis. However, the nano-object to nanomaterial transformation constitutes an unexplored domain where almost every aspect remains to be discovered (see below).

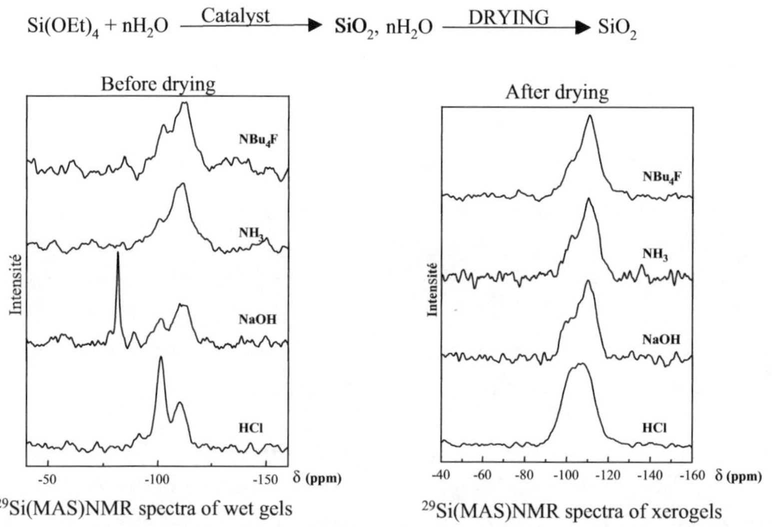

Figure 3.24 Modification of the ^{29}Si (MAS) NMR spectra during the drying of the gel. Reprinted with kind permission from Springer Science and Business Media, ^{29}Si MAS-NMR study of silica gels and xerogels: influence of the catalyst, Framery, E.; Mutin, P.H., *J, Sol-Gel Sci. and Technol., 24,* 191–195. Copyright (2002) Springer

The complete compatibility of silica with any entities coming from molecular chemistry makes *the latter an integral part of materials science*. Consequently, materials science is no longer restricted to methods deriving from thermodynamical control of inorganic compositions. It incorporates all the arsenal of other sectors of chemistry (organic, organometallic, coordination, polymer, biological, etc.). Using inorganic polymerization, *any molecular entities with potentially useful properties can be converted into nanomaterials*. Notice, however, that these important developments imply kinetic control of the matter.

The four types of organic-inorganic hybrid nanomaterials we shall study are:

- *Nanocomposite materials* in which one (or more) molecular entity is incorporated into an inorganic matrix (usually SiO_2) (*cf.* Chapter 4).
- *Grafted materials* in which a nano-object is covalently bonded to a (polymer or oxide) solid matrix (*cf.* Chapter 4).
- *Nanostructured materials*, corresponding to a single phase where molecular entities and the inorganic matrix are permanently linked (*cf.* Chapter 5).
- *Interactive nanomaterials* which are the materials of the future. They involve the coupling of several properties (*cf.* Chapter 6).

We shall also see that the prospects opened up by inorganic polymerization lead to extraordinary potential at the interface of physics and chemistry (*cf.* Chapter 7).

3.7.1 Introduction to the Chemistry of Other Oxides

In the foregoing sections, we discussed mainly SiO_2 which is a very special case.[91] Its polymerization by the sol-gel process is very convenient and very general. A great number of reviews and books have been published on this much studied subject. In addition, the material is relatively inexpensive.

However, this success should not mask other possibilities. The compatibility with molecular chemistry has been observed with several other oxides (Al_2O_3, TiO_2, SnO_2, etc.) obtained by the sol-gel process.[91] The examples are naturally fewer, but are nonetheless significant. It is important to stress that many oxides – in particular the majority of transition metal oxides – have interesting physical properties:* TiO_2 has a photovoltaic effect which can be used to transform light into electricity; SnO_2 is a narrow gap semiconductor; Fe_2O_3, Fe_3O_4 and NiO are magnetic oxides; In_2O_3 is an electric conductor, etc. Given the possibility of incorporating molecular entities possessing usable properties into these matrices, chemists can hope someday to tailor make nanomaterials with interactive properties.

Note, however, that sol-gel processes leading to these oxides have not been extensively studied. Only the preparations of Al_2O_3 are relatively well known. TiO_2 and SnO_2 have been the subjects of some investigations. *Nonetheless, the inorganic polymerization of the majority of oxides to produce hybrid materials remains to be discovered.* Indeed, most methods give precipitates. Each oxide has its own chemical properties which are different from those of silica. For example, the rate of hydrolysis/rate of polycondensation ratio is very different from one metal to the next. It is, therefore, necessary to find the precursors and the experimental conditions for hydrolytic polymerization best adapted to the *formation of a colloidal sol necessary for shaping the material.*

* It is not case for SiO_2 whose only interesting property is a strong dielectric constant, eventually adjustable.

Thus, hydrolytic polymerization of alkoxides, which gives excellent results with SiO_2, will not give a sol with most of the other elements (Sn, Ti, V, etc.). In fact, for the majority of metals, simple functional groups (*e.g.* halogens, alkoxides) lead to precipitates which do not permit the control of the inorganic polymerization. To gain this control, fundamental studies are necessary to find the precursors of sol-gel polymerizations and to develop new inorganic matrices compatible with molecular entities. This chemistry is barely touched upon, but represents an important challenge for the discovery of new materials. *It is important to note that coordination chemistry can be an abundant source of new precursors.*

3.7.2 Generalization to Other Types of Combinations

In addition to the oxides, there exist other combinations like nitrides, sulfides or phosphides. Preparation of this type of combination by inorganic polymerization *remains to be discovered, from both the chemical point of view and that of the potential properties* of the solids obtained. Some methods for preparing nitrides have been studied. Some 'more sophisticated' systems with mixed functions have also been investigated.

A search for nitride-based ceramics has brought about polymerization experiments in which oxygen is replaced by nitrogen. Bradley and colleagues[92–95] have described the first gel obtained by indirect ammonolysis of $SiCl_4$.[96]

The precursor (**I**) is prepared from $SiCl_4$.[96] When (**I**) is treated with a catalytic amount of trifluoromethanesulfonic acid, an auto-transamination occurs with elimination of Me_2NH and formation of the cyclic trimer (**II**). This compound reacts with NH_3 to give the translucent gel $[Si(NH)_{2n}(NMe_2)_{4-2n}]$ (Figure 3.25).

By reaction of bistrimethylsilylcarbodiimide ($Me_3Si-N=C=N-SiMe_3$) with $SiCl_4$, Riedel obtained a solid of composition $[Si(N=C=N)_2]_n$. This solid has the characteristics of an amorphous polycondensed system. It is the analog of silica where the carbodiimide group has replaced the oxygen atom. This work has been extended to some organic-inorganic hybrid systems and to borazines (Figure 3.26).[97]

At present, these are only preliminary works. These successes show, however, that there exist domains opening up for completely original combinations whose exploration may prove fruitful.

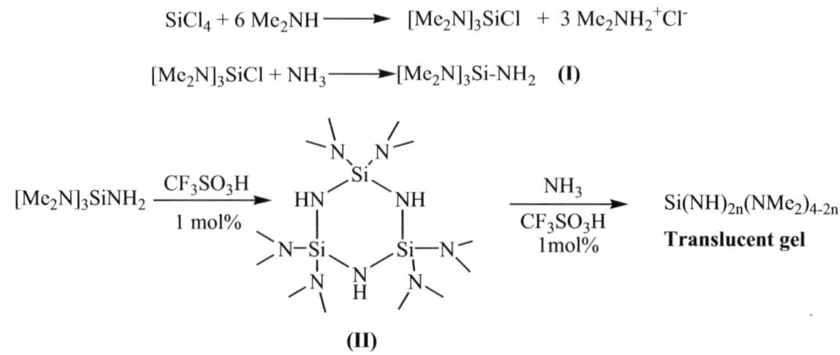

Figure 3.25 Composition of the $Si(NH)_{2n}(NMe_2)_{4-2n}$ translucent gel. Reprinted with permission from *Angew. Chem., Int. Ed.*, Non-oxide sol-gel chemistry: preparation from tris(dialkylamino)silazanes of a carbon-free, porous, silicon diimide gel, Rovai, Ricardo; Lehmann, Christian W.; Bradley, John S., *38*, 2036–2038. Copyright (1999) Wiley-VCH

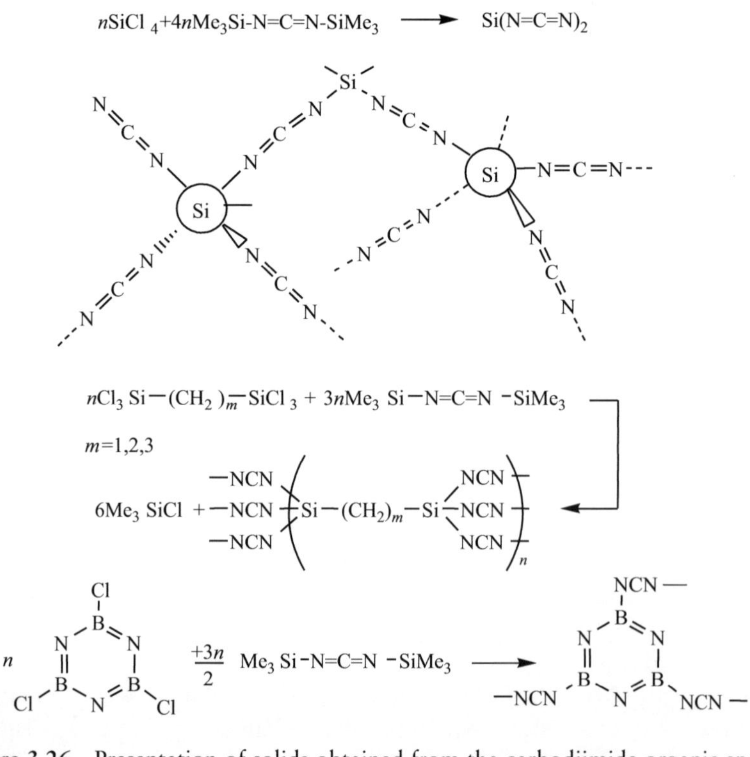

Figure 3.26 Presentation of solids obtained from the carbodiimide organic spacer

REFERENCES

[1] D.J.M. Bevan, P. Hagenmuller, *Nonstoichiometric Compounds: Tungsten Bronzes, Vanadium Bronzes and Related Compounds*, Pergamon, Elmsford, **1975**.

[2] C. Kittel, *Introduction to Solid State Physics*, 5th Edn, John Wiley & Sons, Ltd, New York, **1976**.

[3] D.W. Bruce, D. O'Hare, *Inorganic Materials*, 2nd Edn, John Wiley & Sons, Ltd, New York, **1997**.

[4] W.D. Callister,(ed), *Materials Science and Engineering, An Introduction*, 4th Edn, John Wiley & Sons, Ltd, Chichester, **1996**.

[5] J. Ferraris, D.O. Cowan, V. Walatka, J.H. Perlstein, *J. Am. Chem. Soc.* **1973**, *95*, 948.

[6] D. Jerome, A. Mazaud, M. Ribault, K. Bechgaard, *J. Phys. Lett.* **1980**, *41*, 95.

[7] L. Valade, D. de Caro, M. Basso-Bert, I. Malfant, C. Faulmann, B. Garreau de Bonneval, J.-P. Legros, *Coord. Chem. Rev.* **2005**, *249*, 1986and references therein.

[8] D. de Caro, J. Fraxedas, C. Faulmann, I. Malfant, J. Milon, J.-F. Lamère, V. Collière, L. Valade, *Adv. Mater.* **2004**, *16*, 835.

[9] L. Brossard, M. Ribault, M. Bousseau, L. Valade, P. Cassoux, *C. R. Acad. Sci. Paris* **1986**, *302*, 205.

[10] P. Cassoux, L. Valade, Molecular *Inorganic Superconductors*, D.W. Bruce, D. O'Hare (eds), John Wiley & Sons, Ltd, Chichester, **1992**, 1–58.

[11] L. Ouahab, P. Batail, C. Perrin, C. Garrigou-Lagrange, *Mater. Res. Bull.* **1986**, *21*, 1223.

[12] P. Batail, K. Boubekeur, M. Fourmigue, A. Dolbecq, J.-C. Gabriel, A. Guirauden, C. Livage, S. Uriel, *New. J. Chem.* **1994**, *18*, 999.

[13] J.-C. Gabriel, K. Boubekeur, S. Uriel, P. Batail, *Chem. Rev.* **2001**, *101*, 2037.

[14] P. Batail, *Chem. Rev.* **2004**, *104*, 4887.

[15] P. Rabu, M. Drillon, *Adv. Eng. Mater.* **2003**, *5*, 189.

[16] E. Philippot, D. Palmier, M. Pintard, A. Goiffon, *J. Solid State Chem.* **1996**, *123*, 1.

[17] J. Detaint, B. Capelle, A. Zarka, D. Palmier, E. Philippot, *Ann. Chim.* **1997**, *22*, 651.

[18] G. Ertl, H. Knoezinger, J. Weitkamp (eds), *Handbook of Heterogeneous Catalysis 5*, Wiley-VCH, Weinheim, **1997**.

[19] J.L. Atwood, J.E.D. Davies, D.D. MacNicol, F. Vogtle, J.M. Lehn,(eds), *Comprehensive Supramolecular Chemistry*, Elsevier Science Ltd, Oxford, **1996**.

[20] H. van Bekkum, E.M. Flanigen, P.A. Jacobs, J.C. Jansen,(eds), *Introduction to Zeolite Science and Practice*, 2nd Edn, Stud. Surf. Sci. Catal. 137, Elsevier, Amsterdam, **2001**.

[21] F. Di Renzo, F. Fajula, *Stud. Surf. Sci. Catal.* **2005**, *157*, 1.

[22] H. Robson, K.P. Lillerud,(eds), *Verified Syntheses of Zeolitic Materials*, Elsevier, Amsterdam, **2001**.

[23] J.A. Rabo, M.W. Schoonover, *Appl. Catal. A* **2001**, *222*, 261.

[24] Ch. Baerlocher, W.M. Meier, D.H. Olson, *Atlas of Zeolite Framework Types*, 5th Edn, Elsevier, Amsterdam, **2001**.

[25] G. Ferey, *Chem. Mater.* **2001**, *13*, 3084.

[26] C. Mellot-Draznieks, J.M. Newsam, A.M. Gorman, C.M. Freeman, G. Ferey, *Angew. Chem., Int. Ed.* **2000**, *39*, 2270.

[27] G. Ferey, *J. Solid State Chem.* **2000**, *152*, 37.

[28] S. Girard, A. Tuel, C. Mellot-Draznieks, G. Ferey, *Angew. Chem., Int. Ed.* **2002**, *41*, 972.

[29] G. Ferey, *La Recherche* **2002**, *357*, 38.

[30] G. Ferey, C. Mellot-Draznieks, C. Serre, F. Millange, J. Dutour, S. Surble, I. Margiolaki, *Science* **2005**, *309*, 2040.

[31] G. Ferey, Personal communication.
[32] M. Birot, J.-P. Pillot, J. Dunoguès, *Chem. Rev.* **1995**, *95*, 1443.
[33] S. Yajima, J. Hayashi, M. Omori, Japanese Appl. *7550.223*, **1975**; *Chem. Abstr.* **1977**, *86*, 30940.
[34] S. Yajima, M. Omori, J. Hayashi, Japanese Appl. *7550.529*, **1975**; *Chem. Abstr.* **1977**, *86*, 45291.
[35] W. Verbeek, G. Winter,German Patent 2.236.078, **1974**; *Chem. Abstr.* **1974**, *81*, 50911.
[36] R.J.P. Corriu, P. Gerbier, C. Guérin, B.J.L. Henner, A. Jean, P.H. Mutin, *Organometallics* **1992**, *11*, 2507.
[37] R.J.P. Corriu, P. Gerbier, C. Guérin, B.J.L. Henner, *Angew. Chem., Int. Ed. Engl.* **1992**, *31*, 1195.
[38] R.J.P. Corriu, P. Gerbier, C. Guérin, B.J.L. Henner, *Adv. Mater.* **1993**, *5*, 380.
[39] R.J.P. Corriu, P. Gerbier, C. Guérin, B.J.L. Henner, *Chem. Mater.* **2000**, *12*, 805.
[40] R.J.P. Corriu, P. Gerbier, C. Guérin, B.J.L. Henner, *J. Mater. Chem.* **2000**, *10*, 2173.
[41] R.J.P. Corriu, *Angew. Chem., Int. Ed.* **2000**, *39*, 1376.
[42] M. Ebelmann, *Ann. Chim. Phys.* **1846**, *16*, 129.
[43] M. Ebelmann, *C. R. Acad. Sci. Paris* **1847**, *25*, 854.
[44] L.L. Hench, J.K. West, *Chem. Rev.* **1990**, *90*, 33.
[45] C.J. Brinker, G.W. Scherer, *Sol-Gel Science: The Physics and Chemistry of Sol-Gel Processing*, Academic Press, San Diego, **1990**.
[46] J. Chojnowki, M. Cypryk, K. Kazmierski, K. Rózga, *J. Non-Cryst. Solids* **1990**, *125*, 40.
[47] N. Auner, J. Weis,(eds), *Organosilicon Chemistry V: From Molecules to Materials*, Wiley-VCH, Weinheim, **2003**.
[48] N. Auner, J. Weis,(eds), *Organosilicon Chemistry IV: From Molecules to Materials*, Wiley-VCH, Weinheim, **2000**.
[49] N. Auner, J. Weis,(eds), *Organosilicon Chemistry III: From Molecules to Materials*, Wiley-VCH, Weinheim, **1998**.
[50] N. Auner, J. Weis,(eds), *Organosilicon Chemistry II: From Molecules to Materials*, Wiley-VCH, Weinheim, **1996**.
[51] M.A. Brook, *Silicon in Organic, Organometallic, and Polymer Chemistry*, John Wiley & Sons, Ltd, New York, **2000**.
[52] Z. Rappoport, Y. Apeloig,(eds), *The Chemistry of Organic Silicon Compounds*, Volume 2, John Wiley & Sons, Ltd, New York, **1998**.
[53] P. Gomez-Romero, C. Sanchez, (eds), *Functional Hybrid Materials*, Wiley-VCH, Weinheim, **2004**.
[54] R. Corriu, R. Perz, C. Reyé, *Tetrahedron* **1983**, *39*, 999.
[55] S.J. Gregg, S.W. Sing, *Adsorption, Surface Area and Porosity*, Academic Press, London, **1982**.
[56] S. Brunauer, P.H. Emmett, E. Teller, *J. Am. Chem. Soc.* **1938**, *60*, 309.
[57] E. Barrett, L.G. Joyner, P.P. Halenda, *J. Am. Chem. Soc.* **1951**, *73*, 373.
[58] E. Framery, P.H. Mutin, *J. Sol-Gel Sci. Technol.* **2002**, *24*, 191.
[59] R.J.P. Corriu, J. Moreau, P. Thepot, M. Wong Chi Man, *Chem. Mater.* **1992**, *4*, 1217.
[60] R.J.P. Corriu, J. Moreau, P. Thepot, M. Wong Chi Man, *J. Mater. Chem.* **1994**, *4*, 987.
[61] B.M. Baugher, D.A. Loy, S. Prabakar, R.A. Assink, K.J. Shea, H. Oviatt, *Mater. Res. Soc. Symp. Proc.* **1995**, *371*, 253.
[62] D.A. Loy, B.M. Baugher, S. Prabakar, R.A. Assink, K.J. Shea, *Mater. Res. Soc. Symp. Proc.* **1995**, *371*, 229.

[63] J.-P. Bezombes, C. Chuit, R.J.P. Corriu, C. Reye, *J. Mater. Chem.* **1998**, *8*, 1749.
[64] J.-P. Bezombes, C. Chuit, R.J.P. Corriu, C. Reye, *J. Mater. Chem.* **1999**, *9*, 1727.
[65] G. Cerveau, R.J.P. Corriu, C. Fischmeister-Lepeytre, *J. Mater. Chem.* **1999**, *9*, 1149.
[66] G. Cerveau, R.J.P. Corriu, E. Framery, *J. Mater. Chem.* **2000**, *10*, 1617.
[67] G. Cerveau, R.J.P. Corriu, E. Framery, *Polyhedron* **2000**, *19*, 307.
[68] R.J.P. Corriu, J. Moreau, M. Wong Chi Man, *J. Sol-Gel Sci. Technol.* **1994**, *2*, 87.
[69] G. Cerveau, R.J.P. Corriu, C. Lepeytre, *J. Organomet. Chem.* **1997**, *548*, 99.
[70] R.J.P. Corriu, J. Moreau, P. Thepot, M. Wong Chi Man, C. Chorro, J.-P. Lère-Porte, J.-L. Sauvajol, *Chem. Mater.* **1994**, *6*, 640.
[71] G. Cerveau, R.J.P. Corriu, C. Lepeytre, P.H. Mutin, *J. Mater. Chem.* **1998**, *8*, 2707.
[72] C. Lepeytre,Thesis, Montpellier, **1998**.
[73] R.J.P. Corriu, *Polyhedron* **1998**, *17*, 925.
[74] R.J.P. Corriu, J. Moreau, P. Thepot, M. Wong Chi Man C. Chorro, J.-P. Lère-Porte, J.-L. Sauvajol, *Chem. Mater.* **1994**, *6*, 640.
[75] G. Cerveau, R.J.P. Corriu, E. Framery, *Chem. Commun.* **1999**, 2081.
[76] G. Dubois, C. Reye, R.J.P. Corriu, S. Brandes, F. Denat, R. Guilard, *Chem. Commun.* **1999**, 2283.
[77] G. Cerveau, R.J.P. Corriu, E. Framery, *J. Mater. Chem.* **2001**, *11*, 713.
[78] G. Cerveau, R.J.P. Corriu, E. Framery, *C. R. Acad. Sci.* **2001**, *4*, 79.
[79] G. Cerveau, R.J.P. Corriu, E. Framery, F. Lerouge, *Chem. Mater.* **2004**, *16*, 3794.
[80] G. Cerveau, R.J.P. Corriu, E. Framery, F. Lerouge, *J. Mater. Chem.* **2004**, *14*, 3019.
[81] F. Lerouge, G. Cerveau, R.J.P. Corriu, *J. Mater. Chem.* **2006**, *16*, 90.
[82] N. Bellec, F. Lerouge, O. Jeannin, G. Cerveau, R.J.P. Corriu, D. Lorcy, *J. Organomet. Chem.* **2006**, *691*, 5794.
[83] F. Lerouge, G. Cerveau, R.J.P. Corriu, *New J. Chem.* **2006**, *30*, 1364.
[84] F. Lerouge, G. Cerveau, R.J.P. Corriu, C. Stern, R. Guilard, *Chem. Commun.* **2007**, 1553.
[85] G. Cerveau, R.J.P. Corriu, E. Framery, S. Ghosh, P.H. Mutin, *J. Mater. Chem.* **2002**, *12*, 3021.
[86] H. Marsmann, *^{29}Si NMR Spectroscopic Results*, P. Diehl, E. Fluck, R. Kosfeld,(eds), Springer Verlag, Berlin, **1981**.
[87] E.A. Williams, *The Chemistry of Organosilicon Compounds*, S. Pataï, Z. Rappoport, (eds), John Wiley & Sons, Ltd, New York, **1989**, 511.
[88] M. Magi, E. Lippmaa, A. Samoson, G. Engelhardt, A.R. Grimmer, *J. Phys. Chem.* **1984**, *88*, 1518.
[89] K.J. Shea, D.A. Loy, O.W. Webster, *J. Am. Chem. Soc.* **1992**, *114*, 6700.
[90] H.W. Oviatt, K.J. Shea, J.H. Small, *Chem. Mater.* **1993**, *5*, 943.
[91] J.-P. Jolivet, *De la solution à l'oxyde*, Inter Editions, Paris, **1994**.
[92] F. Cheng, S.M. Kelly, F. Lefebvre, S. Clark, R. Supplit, J.S. Bradley, *J. Mater. Chem.* **2005**, *15*, 772.
[93] F. Cheng, S.J. Archibald, S. Clark, B. Toury, S.M. Kelly, J.S. Bradley, F. Lefebvre, *Chem. Mater.* **2003**, *15*, 4651.
[94] F. Cheng, B. Toury, F. Lefebvre, J.S. Bradley, *Chem. Commun.* **2003**, 242.
[95] D. Farrusseng, K. Schlichte, B. Spliethoff, A. Wingen, S. Kaskel, J.S. Bradley, F. Schuth, *Angew. Chem., Int. Ed.* **2001**, *40*, 4204.
[96] R. Rovai, C.W. Lehmann, J.S. Bradley, *Angew. Chem., Int. Ed.* **1999**, *38*, 2036.
[97] R. Riedel, E. Kroke, A. Greiner, A.O. Gabriel, L. Ruwisch, J. Nicolich, P. Kroll, *Chem. Mater.* **1998**, *10*, 2964.

4

From Nano-Object to Nanomaterial

We have presented nano-objects and drawn attention to their importance in the exploration of the nanoworld, which can lead to entirely original physical properties, resulting in a new generation of materials. However exciting, innovative and fruitful this may be, this development is only a part of nanosciences. For chemists, the main objective of nanosciences is the discovery of new preparations of nanomaterials by the bottom-up approach. A nano-object is not a material, only a chemical product with a certain property. *Without assemblage and shaping, it remains a laboratory curiosity.*

The purpose of this chapter is to present some different types of nanomaterials currently available and to describe their likely development in the near future.

4.1 THE DIFFERENT TYPES OF NANOMATERIALS

What do we mean by nanomaterials? As with any materials, nanomaterials are stable solids exhibiting usable (physical, chemical, mechanical, even biological) properties. They must be shaped, *i.e.* they must come as films, matrices or fibers. In other words, they must present specifications well defined for a precise application. However, nanomaterials differ from classical materials on several points. First, molecular (organic, organometallic, coordination, etc.) chemistry is extensively involved. Secondly, these materials are devised from the elementary scale (molecule, aggregate, cluster, etc.). Indeed, the nano-object is conceived and synthesized in order to produce a material having clearly defined properties. Their synthesis (molecule, aggregate, cluster, etc.), nevertheless, does not require new chemistry; existing knowledge is sufficient.

However, and this is the third difference, *the transformation of the nano-object into a nanomaterial will not be done* by existing thermal methods (*cf.* Chapter 3). In effect, the use of temperatures exceeding 400 °C is not compatible with the organic entities frequently present in nano-objects. Nanomaterials will be produced either by (*organic, inorganic or supramolecular*) *polymerization or by hydrothermal synthesis.* All these methods, which respect the nano-object's structure, are under kinetic control, while the classical thermal methods of material sciences are under thermodynamic control. Some of the methods have been used, but they have their limitations. Fortunately, the different methods of organic and inorganic polymerization open up vast opportunities for discovery. These will gradually come to light with the problems discussed. We shall see that *a major part of the chemistry permitting access to most of the nanomaterials remains to be discovered.*

Finally, let us recall that the preparation of nanomaterials from nano-objects involves two independent but inseparable objectives: assemblage and shaping. On the one hand, the nano-object must be transformed by chemical means (assemblage or incorporation into a matrix) into a solid material with the expected properties. On the other hand, it is necessary to obtain a material, which can be *shaped into a form adapted to the intended application* (coating, film, fiber, matrix or a solid with controlled porosity).

Having mentioned the synthesis of nano-objects, we shall now examine the different types of existing nanomaterials and present their main characteristics as well as the methods presently used for their preparation. We shall also try to outline some prospects for their development.

There are four types of nanomaterials (Figure 4.1), which differ from each other by the method of incorporating the nano-object into the matrix:

(1) *Nanocomposites* correspond to the inclusion of a nano-object in an organic or inorganic (generally an oxide, see below) matrix. They are biphasic systems, since the matrix and the nano-object can be separated.
(2) In *grafted nanomaterials,* a nano-object is linked to a solid surface by a covalent bond. The solid surface can be an organic polymer or, most often, an inorganic oxide (see below).

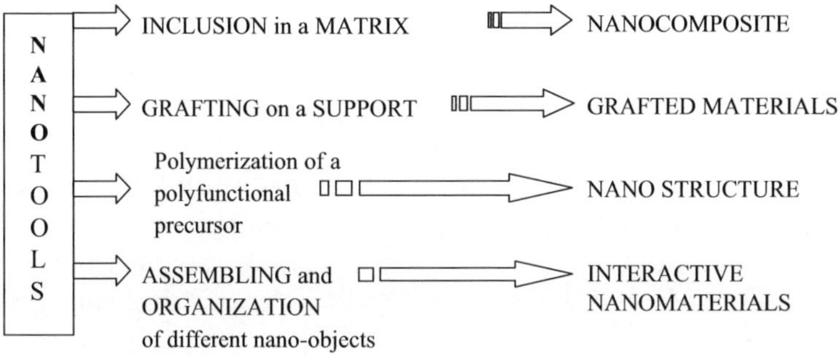

Figure 4.1 Schematic diagram of the different types of nanomaterials. Reproduced by permission of *L'actualité chimique*

(3) *Nanostructures* come as a single phase with the nano-object integrated into an inorganic matrix. These materials will be discussed separately in Chapter 5.
(4) *Interactive nanomaterials* are the beginning of nanomaterials of the future. They correspond to nanomaterials in which two groups inducing different properties are included in the materials and are precisely located at the nanometric scale* (*cf.* Chapter 6).

In this chapter we shall examine the nanocomposites and the grafted nanomaterials, which have been studied for several decades and have the most promising applications, including organic synthesis. Nanostructures and interactive nanomaterials will be discussed in Chapters 5 and 6, respectively.

4.2 INORGANIC POLYMERIZATION: A MAJOR ROUTE TO NANOMATERIALS

Before considering composite nanomaterials, we must emphasize the importance of inorganic polymerization in material science. Sol-gel processes give access, by *kinetically controlled methods*, to many inorganic combinations, with oxides up to now being the most important. This methodology represents at present the principal inclusion method for

*'Smart materials' are not included in this list as they belong not to nanosciences but rather to material science and related technologies. They correspond to systems assembled at scales larger than a micrometer. Undoubtedly, nanomaterials can be incorporated into such systems. However, interactive nanomaterials proceed from a completely different, and at present purely exploratory, conception.

preparing nanocomposites. As we have seen, *inorganic polymerization is entirely compatible with all molecular chemistry* (organic, organometallic, chemistry of the main elements, etc.). It is also compatible with coordination chemistry, thus allowing the inclusion in nanomaterials of metal ions (alkali, alkaline earth, heavy and transition metals, lanthanides, actinides, etc.). In addition, it is totally compatible with organic or organometallic polymers. Finally, it permits the stabilization of more complex inorganic combinations, such as oxides or metal nanoparticles (*cf.* Chapter 5).

Moreover, this method allows the shaping of the material (film, fiber and matrix) during the viscous sol step. It will be also possible to optimize the texture characteristics of the material by using the parameters controlling the polymerization kinetics (*cf.* Chapter 5). It is thus possible to prepare entirely new materials 'à la carte', so to speak.

Nanomaterials thus obtained are commonly called organic-inorganic hybrids (and often 'hybrid materials' for short).

4.3 NANOCOMPOSITE MATERIALS

In this chapter, we shall succinctly present nanocomposite materials, which are well known and widely found. These nanomaterials were and still are the subject of many interesting works,[1] since they represent the most convenient way for *transforming a nano-object into a shaped material*. At the present time, they are the most frequently used materials. Nanocomposites are obtained by encasement or inclusion of nano-objects in an organic or inorganic matrix. The organic matrix is usually a polymer easy to shape. Inorganic matrices, of which the most often used is silica, are generally made of oxides and prepared by hydrolytic polycondensation of the sol-gel type.

4.3.1 Nanocomposites in Silica Matrices[2–5]

The fact that SiO_2 is employed in the majority of cases is easy to understand. This oxide can be obtained from molecular precursors produced industrially in large quantities. It has been the subject of numerous studies, which have shown its total compatibility with molecular, even biological, systems. It represents the most convenient and most efficient route, based on a well-established know-how, to prepare and shape materials endowed with a given property.

Introduction of photoluminescent organic molecules in silica matrices has been widely studied. The nanomaterial is shaped during the sol stage into films, fibers or matrices. It is important, however, that the organic entities are separated from each other. Hydrophobic (or rather lipophilic) interactions within the hydrophilic silica during the polymerization stage lead to their spontaneous aggregation. The result is a decrease of the photoluminescence optical yield. A regular distribution becomes possible by addition of an additive less hydrophilic than SiO_2. For example, a precursor bearing a hydrophobic substituent such as $CH_3-Si(OR)_3$ is introduced during the polycondensation of $Si(OR)_4$.

An example illustrating the many possibilities offered by nanocomposites is the inclusion of ion-doped nanoparticles in modified silica matrices. Nanoparticles of oxides (YVO_4 or $LaPO_4$) doped with photoluminescent ions (Eu^{3+}, Co^{3+}, Tb^{3+}, etc.) of controlled sizes have been incorporated into SiO_2 matrices. Additives (citrates, polyphosphates) can be used to regulate their distribution within the bulk material. A large range of emission wavelengths has been obtained by using the doping ion/nanoparticle ratio.[6,7]

Organic polymers have been often used as matrices for the encasement of organic entities. Numerous materials of this type have been studied, in particular for nonlinear optics. The polymer matrices (polyacrylate, polystyrene, polypropylene, etc.) are also suitable for the inclusion of molecular nano-objects.

The choice between organic or inorganic polymers for the matrix is essentially determined by the intended application and the expertise of the experimentalists.

4.3.2 Some Developments of Nanocomposites

A substantial development of nanocomposite materials can be anticipated, considering their convenience.[8–29] The relevant chemistry deals with the nano-object synthesis and its introduction into a silica matrix. The role of the latter is prominent: it is the matrix, which transforms the nano-object into nanomaterial, thus permitting it to display its potential properties.

Until now, the matrix's role has been essentially to allow the shaping of the material: *the nano-object brings the property; the matrix makes it a material.* We can anticipate, however, that the matrix can play a more active role by amplifying or adjusting properties of the material. Let us keep in mind that the properties of the SiO_2 matrix can be modified by molecular additives. For example, alkyl groups lower the hydrophilicity. Incorporation of (NH_2), (SH), (CO_2H), PO_3Et_2, etc. allows the introduction of functional entities (metal nanoparticles or complexes). From what is known about SiO_2 matrices, we can foresee that the possibilities offered by this type of material will be substantially larger.

Matrices capable of coupling their properties with those of the nano-objects can also be envisaged. Here is an outline of some emerging developments.

4.3.3 Presentation of Potential New Matrices

Nanostructured organic-inorganic hybrid matrices are very promising, since they already contain a molecular entity, which may have a property complementary to that of the nano-object. This possibility will be discussed in Chapter 5 (Section 5.7.4.1), following the presentation of nanostructured materials.

4.3.3.1 Matrices different from SiO_2

The development of matrices different from SiO_2 is also to be expected. This has been discussed in Chapter 3 (Section 3.6.2). Clearly, matrices having physical properties, such as TiO_2 (photovoltaic), SnO_2 (semiconducting), Fe_3O_4 or Fe_2O_3 (magnetic), may present interesting couplings with the nano-objects they contain.

To prepare these matrices effectively, inorganic polymerization of oxides having useful physical properties must be mastered. Unfortunately, at present, both the know-how and the fundamental knowledge are lacking and this is the major difficulty for chemists to overcome. This chemistry must be explored, since inclusion of the nano-objects implies perfect control of the hydrolytic polycondensation and of the balance between hydrophilicity and hydrophobicity. Almost all this domain is yet to be discovered (*cf.* Section 3.6.2).

4.3.3.2 Matrices based on vitreous materials

Inorganic glass matrices open up a very interesting subdivision in the field of nanocomposites. Glasses are materials known since antiquity. They have mechanical properties and great stability. They can be shaped, for at temperatures higher than the glass transition temperature (T_g), glasses become plastic and consequently, can be transformed into a fiber, matrix or coating. However, in the majority of cases, T_g is higher than 500 °C, hence incompatible with organic or organometallic entities.

These viscous liquids can behave as matrices and incorporate additives (e.g. glasses containing metal oxides). Colored glasses can thus be produced, some of which are artistic masterpieces. The incorporation of ions has also led to interesting scientific products, for example lanthanide lasers or optical amplifiers for telecommunications using compounds containing Er^{3+} ions. Quantum dots have been produced by the introduction of metal nanoclusters or semiconductors. These nanocombinations present quantum confinement effects or nonlinear optical properties.

More recently, glasses have been prepared from combinations containing the elements B/O/F in various ratios. They have relatively low T_g values, sometimes lower than 300 °C, and are therefore compatible with many organic products or organometallic products or coordination complexes. Fibers, films or matrices can be made. These glass matrices show interesting prospects for the production of nanocomposite materials.[30,31]

Lucas and colleagues have prepared glasses corresponding to mixed lead and tin fluorophosphates.[32] They are obtained at 320 °C from a mixture of powders of defined compositions, for example PbF_2, SnF_2, SnO and P_2O_5. They exhibit a T_g of 220 °C, compatible with the majority of organic entities. Dyes such as rhodamine G have been incorporated into these glasses homogeneously, in order to avoid the autoaggregation of the organic compounds, which would strongly diminish the luminescence. In mixed (Sn–Pb) fluorophosphate matrices, the organic molecules are separated from each other. This permits very intense emissions under radiation as well as an excellent regularity of the signal with the passing of time. All these features promise interesting applications in the laser field.

Note also that these glass matrices can be transformed into microporous solids by controlled foaming. These pores in turn can accept other organic or inorganic entities: it is thus possible to obtain glasses containing two types of entities endowed with physical or chemical properties, which can interact. This is a purely inorganic approach to interactive nanomaterials, which will be discussed in Chapter 6. This domain, which promises a broad range of potentialities, will be widely developed in the near future.

4.4 GRAFTED MATERIALS

4.4.1 Advantages of Solid Supports

Grafting chemical functions on solid supports has been known and used for a long time. In effect, this methodology circumvents tedious separation operations. It increases yield and selectivity by eliminating most of the unwanted reactions, thus favoring the desired reaction. Its main value lies in the fact that it combines the *selectivity* of the grafted group towards the entity to be separated and the *ease* of separation (simple filtration).

The grafting can be realized on an inorganic matrix, which usually is SiO_2. It can also be carried out on an organic polymer (polyacrylate, polyethylene, polystyrene, etc.). The choice of the support depends on the intended objectives. Some examples will be discussed later.

We have discussed (*cf.* Section 1.2) the separation of lead from drinking water using a silica matrix on which is grafted a molecule capable of chelating selectively Pb^{2+}. This method advantageously replaces the extraction by washing, which is much less convenient, much less efficient and above all much less selective. In the purified water, the remaining concentration of lead is $<5\ \mu g\ l^{-1}$: the quasi-totality of Pb is thus chelated and is, in fact, separated. Note that the content of Mg^{2+} and Ca^{2+} remains unchanged. Moreover, the material can be easily regenerated by an acid washing at $pH \approx 4$, which eliminates Pb^{2+}, followed by a second washing to bring the pH back to 7 (*cf.* Section 4.5).

Grafted materials are also very useful for certain syntheses. The best-known example is the peptide synthesis whose success earned Professor Bruce Merrifield the Nobel Prize. It constitutes a model illustrating all the advantages offered by grafting methods.[33,34]

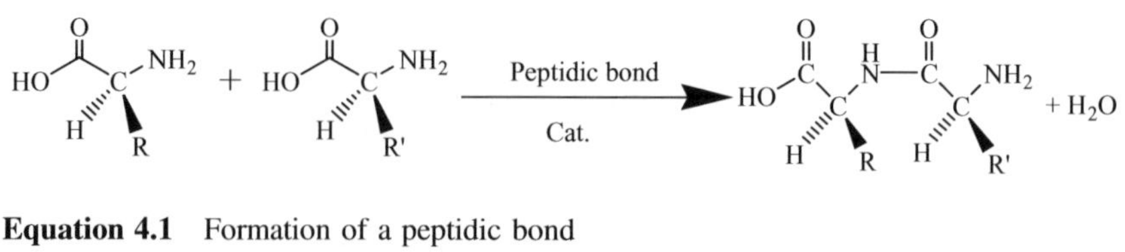

Equation 4.1 Formation of a peptidic bond

Peptides are biological polymers, made of amino acids linked together by the peptidic bond –HN–CO–, which corresponds to the formation of a carbon–nitrogen bond between the NH_2 group of one amino acid and the CO_2H function of another [Equation (4.1)]. Consider a peptidic sequence formed by the amino acids A, B, C, D, E, etc.. The difficulty of the ABCDE peptide synthesis is due to the fact that two amino acids A and B can form two different peptidic bonds: one between the amine group of A and the acid group of B and one between the acid of A and the amine of B. The two dipeptides AB and BA in turn can combine with C to provide four tripeptides CAB, ABC, CBA and BAC and so on, the number of products increasing exponentially with the sequence. Note furthermore that in the case of longer peptides, there exists the possibility of cyclization reactions leading to cycles of different sizes.

The Merrifield method cleverly binds the first amino acid A by its acid function to a solid polymer support bearing NH_2 groups. Thus the only possible reaction of B with the grafted A will occur between the acid of B and the amine of A. Reagent B may be used in excess to improve the yield. Once the reaction is finished, a simple washing suffices to purify the dipeptide attached to the support. In order to avoid secondary reactions, the amino function of B is protected by a group called Boc (tert-butyloxycarbonyl), which later can be easily eliminated by acid hydrolysis with regeneration of the NH_2 group (protection–deprotection sequence). The dipeptide AB is afterwards treated by the amino acid C whose amine function is protected, etc.

Each step: A → AB → ABC → ..., etc. will be executed separately and in each step the amine function remains unprotected. At the end of the sequence, the expected peptide ABCDE is recovered by simple cleavage of the peptidic bond linking the first amino acid A to the support (Figure 4.2).

The method is sufficiently reliable to permit the automation of the synthesis. Thus the synthesis of peptides containing a hundred amino acids has been realized with perfect control of the stereochemistry at every step.

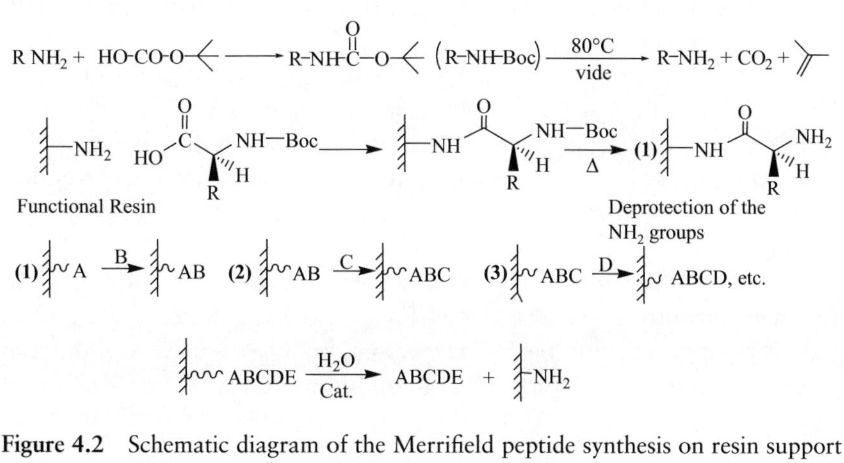

Figure 4.2 Schematic diagram of the Merrifield peptide synthesis on resin support

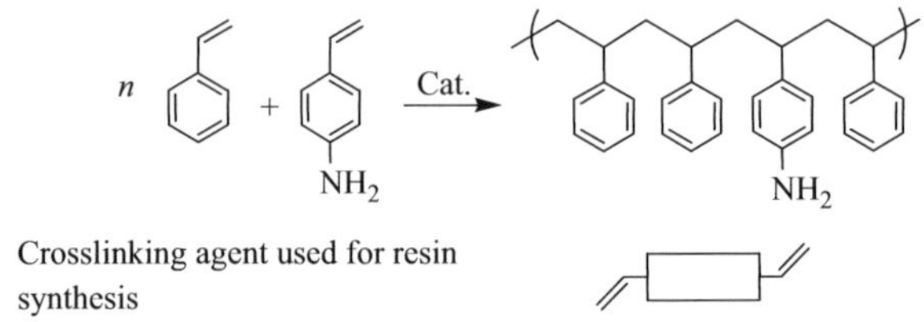

Figure 4.3 Schematic diagram of Merrifield resin preparation

The supports used must have good mechanical performance, which is at present achieved by crosslinking. Thus, Merrifield resins (polystyrenes bearing amino groups) are obtained by copolymerization of styrene and *p*-aminostyrene in the presence of crosslinking reagents which are organic molecules bearing several vinyl groups (Figure 4.3).

4.4.2 General Remarks

Grafted materials are the simplest and most accessible form of nanomaterials, obtained by simple attachment of a nano-object to an organic polymer matrix or an oxide matrix (most frequently SiO_2, for reasons of cost, accessibility, uses and above all for the numerous usable grafts). This attachment is generally made using one or several covalent bonds to link the matrix to a precursor. The precursor comprises a nano-object and a grafting group bound together by a spacer.

The most frequently used organic polymers are polyacrylates, polyacrylamides and polystyrenes. They are obtained by a copolymerization analogous to that shown in Figure 4.3. The role of the support consists mainly in fastening the chosen functionalities in a regular fashion, in permitting an optimal accessibility and in ensuring a reproducible experimental convenience.

Figure 4.4 shows some examples of nano-objects grafted on SiO_2. Grafting on a silica and more generally on an oxide (TiO_2, SnO_2, etc.) can be made, in the majority of cases, through a $Si(OR)_3$ group. Introduction of $Si(OR)_3$ groups on an organic molecule is fairly easy and has been the subject of many works. Some examples will be presented (*cf.* Section 5.2.3).

The most frequently used spacers are aliphatic [generally $(CH_2)_3$]. Aromatic groups (C_6H_4, $C_6H_4CH_2$) are also used. Indeed, the silicon industry supplies a number of precursors, which are used in industrial preparations for attaching silicon-containing units to various solid supports.

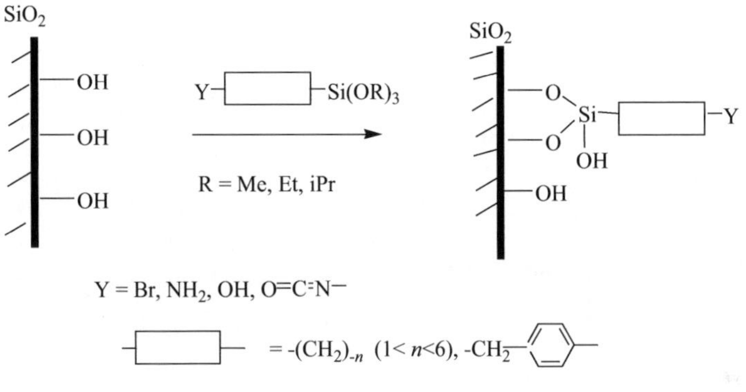

Figure 4.4 Spacers most frequently grafted on SiO_2

4.5 SELECTIVE SEPARATION

One of the major challenges chemists have to face consists of preparing materials for the specific isolation from a mixture of one or several components, either unwelcome (pollution) or, in contrast, desirable (enrichment). This concerns inorganic ions in solution, but also metallic particles, inorganic oxides, molecules, organic aggregates or even biological entities in effluents. The fundamentals already exist. Molecular chemistry permits the synthesis of selective chelatants of particles, molecules or ions. Coordination chemists and supramolecular chemists have clarified the interactions between chelating and chelated systems.[35–37] The chemistry of organic polymers and inorganic polymerization allow the incorporation of these chelating functions into a number of solid materials. Note that efficient materials for selective separation are critical for depollution problems as well as for important industrial activities such as the electronic or nuclear industries.

Four examples can be given by way of illustration.[38–44] The first one concerns perfecting a cartridge for the depollution of lead in potable waters[39–41] and the second one the treatment of effluents from nuclear power stations before rejection.[42–44] In these applications, the nano-object is a cyclam functionally optimized for a selective chelation. Thus, $(-CH_2-CO-NH_2)$ functions are the obvious choice for chelating lead. For the chelation of actinides ($Ac^{n+}nX^-$), the best results are obtained with $(CH_2-CO_2^-)$ groups.[42–44] Note, however, that other parameters must be considered in order to obtain the material best suited to the intended application.

Let us first take the example of the depollution of lead in drinking water (Figure 4.5).[39–41] The cartridge, planned for common domestic use, must

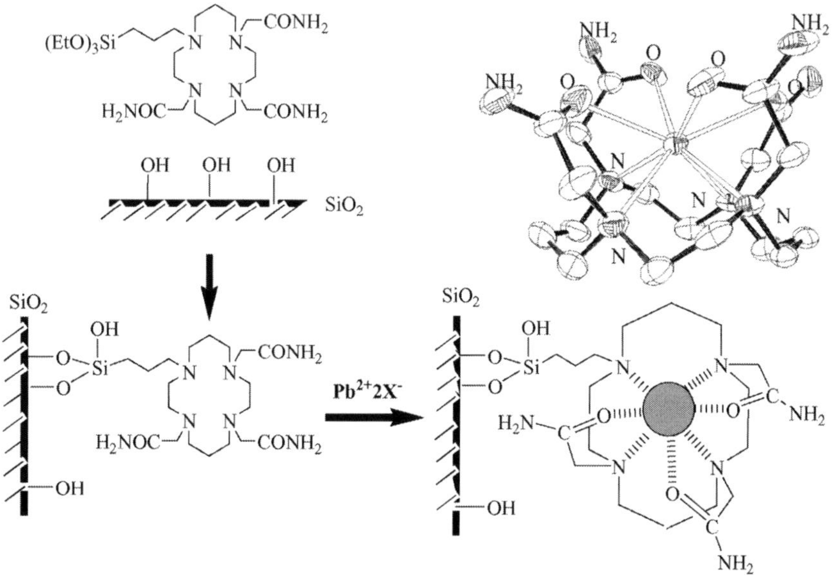

Figure 4.5 Separation of Pb^{2+} in drinking water by using a tetra-azamacrocycle substituted by amido groups and grafted on a silica surface. Reproduced with permission from *Journal of Molecular Liquids*, A molecular approach to remove lead from drinking water by Francois Cuenot, Michel Meyer, Arnaud Bucaille and Roger Guilard, *118*, 1–3, 89–99. Copyright (2005) Elsevier Ltd

be recyclable and the support must be chosen accordingly. SiO_2 is the best choice, as it is very stable and its hydrophilicity permits numerous washings in various conditions. Consequently, the nano-object is functionalized by a $Si(OR)_3$ group on one nitrogen atom through a spacer. This material is afterwards shaped as a cartridge. The concentration of Pb^{2+} in the purified water is $<5\,\mu g\,l^{-1}$ while those of Ca^{2+} and Mg^{2+} are not modified. After saturation with Pb^{2+}, the cartridge is regenerated by a simple washing at $pH \approx 4$ which eliminates Pb^{2+} by protonation of the nitrogen. After washing with water, the cartridge can be reused.

Carboxylate chelatants (instead of amide chelatants) are used for the purification of effluents from nuclear power stations (Figure 4.6). The support must also be different in order to satisfy some unique specifications. On the one hand, the rejected water must be the least radioactive possible; on the other hand, at the end of the process, the radioactive waste must be eliminated. This is the case for Am^3, Pu^{3+} and Np^{3+} wastes in particular, since they are noxious, cannot be reused and, consequently, must be eliminated. Hence, the material must first permit the complete separation of the actinides and secondly, occupy at the end a minimal volume in order to facilitate waste storage. Therefore, an organic polymer

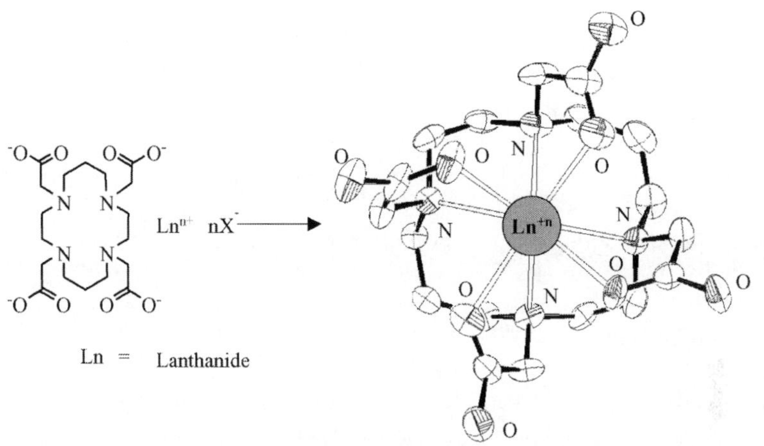

Figure 4.6 X-ray structure of a lanthanide complexed in a functionalized tetra-azamacrocycle. Reproduced with permission from *Coordination Chemistry Reviews*, Conformations and coordination schemes of carboxylate and carbamoyl derivatives of the tetraazamacrocycles cyclen and cyclam, and the relation to their protonation states. Meyer, M.; Dahaoui-Gindrey, V.; Lecomte, C.; Guilard, R.,*178–180*(Pt. 2), 1313–1405. Copyright (1998) Elsevier

matrix was chosen whose combustion leaves no residue, in contrast to an inorganic matrix. The only residues will be the oxides of the used actinides. Once processed, the rejected effluents present radioactivities of $<5\,Bq\,m^{-3}$. *This corresponds to the detection limit and is largely inferior to the natural radioactivity.*

Another beautiful example of selective separation is that of the radioactive Cs^+ generated during uranium fission. This cation has a high mobility in aqueous solution; it is only weakly hydrated because its large size imparts a weak charge density on its surface. Classical membranes cannot separate it from the other alkali ions. As Cs^+ must be selectively separated in the presence of a large quantity of sodium and potassium salts, very efficient methods must be found. This problem has been solved[45–48] using formophenolic resins obtained by copolymerization of calix resorcinarene and formaldehyde. The selectivity is remarkable: some ppm of Cs^+ can be separated from a mixture very rich in Na^+ and K^+ (Figure 4.7).[46–48]

The fourth example is one of the most spectacular. It concerns the *chemical separation* of isotopes.[49–52] Isotopic enrichment (e.g. $^{235}U/^{238}U$ enrichment) is of crucial importance in the nuclear energy domain. The industrial processes used at present are costly in both terms of energy and installations. They use diffusion through membranes or a powerful laser. The chemical route is much less expensive and the separation rates are at least comparable.

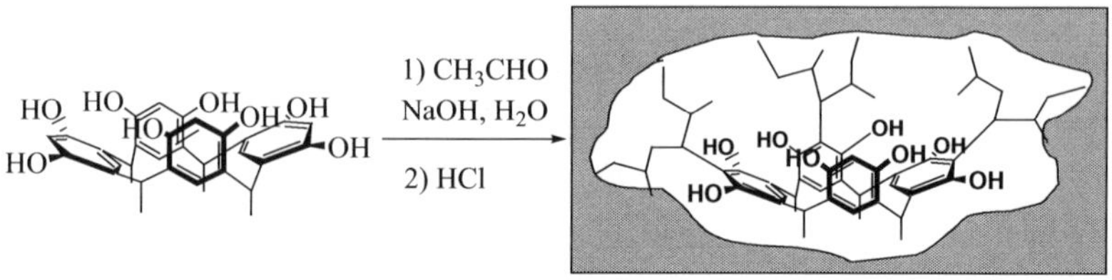

Figure 4.7 Formation of a formophenolic resin having calixarene cavities allowing the selective separation of Cs^+. Reproduced by permission of *L'actualité chimique*

As an example, we can mention the use of selective chelatants fastened on a Merrifield-type polymer matrix to make $^{160}Gd/^{155}Gd$ and $^{65}Cu/^{63}Cu$ enrichments (Figure 4.8).† The separation rate is between 0.06% and 0.16% for each cycle. The list of isotopes *already separated* shows that the method is applicable to most cases.

These few examples demonstrate the possible impact of chemistry in these important and sensitive domains of pollution and nuclear wastes. They illustrate the versatility and adaptability of chemical methods well. Selective chelatants grafted on appropriate supports constitute a very efficient solution to separation problems.

4.6 MATERIALS OBTAINED BY POLYCONDENSATION OF MONOSUBSTITUTED TRIALKOXYSILANES

These are the simplest hybrid materials, obtained by hydrolytic polycondensation of $R{-}Si(OR')_3$ groups [Equation (4.2)].

$$R{-}Si(OR')_3 \xrightarrow[\text{Cat.}]{H_2O} R\text{-}SiO_{1.5} \quad (R' = Me, Et) \quad (Cat. = F^-, H^+, OH^-)$$

Equation 4.2 Synthesis of silsesquioxane materials

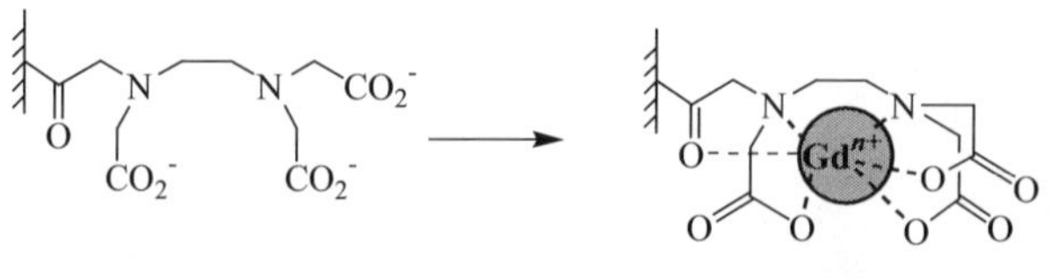

Figure 4.8 Selective separation of $^{160}Gd/^{155}Gd$ on a Merrifield resin grafted with chelating units for the separation of Gd^{3+} isotopes

†The isotopic separation of uranides is classified as 'Defense secret', so the corresponding patents have not been published.

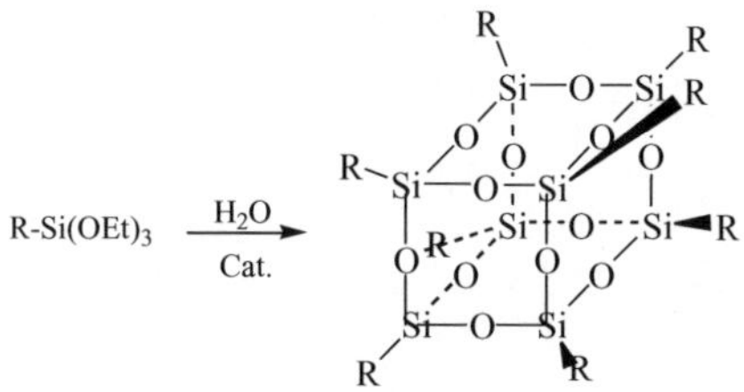

R = H, nPr, nBu, tBu, octyl, decyl, C_6H_5, $C_6H_5CH_2$, etc.

Figure 4.9 Crystallized octasilsesquioxanes with various substituents on silicon

Depending on experimental conditions, different types of materials have been obtained. For example, crystals presenting cubic geometries have been synthesized and studied (Figure 4.9).[53,54]

Many organic R groups have been used and the characteristics of the final material depend on the experimental conditions. In general, resins or systems with controlled porosity are obtained in standard conditions. Table 4.1 shows some examples of these materials.[55,56]

The distribution of organic entities in amorphous materials has been studied with TOF SIMS (time of flight secondary ion mass spectrometry).[55,56] In this method, a material is bombarded with Ga^+ or In^+ ions which liberates the chemical entities on the surface. The ions eliminated are analyzed by mass spectrometry.

This study has shown unambiguously that the R groups are always located on the surface, since it is always possible to detect the fragment R accompanied by secondary fragments due to its scission. For example, in the $ICH_2CH_2CH_2$ $SiO_{1.5}$ case, the fragments I(127) CH_2I(141) ISi(155) C_3H_8I(171) can be detected.

With this method, a great variety of silica having chemical functions on the surface can be prepared.

Table 4.1 Examples of silsesquioxane materials

$R\text{-}Si(OR')_3 \xrightarrow[\text{Cat.}]{H_2O} R\text{-}SiO_{1.5}$
$R = CH_3(CH)_2)_n$ ($n = 0, 1, 2, 3, 4, 5, 7, 9, 11, 15, 17$)
R = 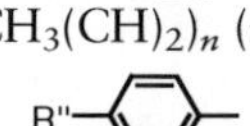 (R'' = H, CH_2-NMe_2, CH_2I, Cl)
R = 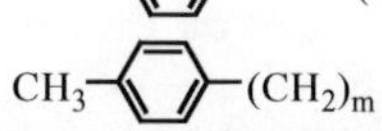 m = 1, 2
R = vinyl, cyclohexyl, tbutyl, $(CH_2)_3$ NH_2, $(CH_2)_3I$, $(CH_2)_3$ NCO, ferrocenyl
$R = (C_6H_5)_2$ P CH_2CH_2–

4.7 MULTISTAGE SYNTHESES – CASCADE REACTIONS

The foregoing examples have illustrated the efficiency of grafted materials in separation problems. The following example shows an application to chemical synthesis.

Multistage synthesis, widely used in organic chemistry, is seriously handicapped by the numerous separations and purifications required in the intermediate stages. These operations are time-consuming and decrease the yields. Industrially, this means an important increase in costs.

A heterogeneous phase can prove to be very useful since, as we have seen, some separations can be carried out by simple filtration. It also has other advantages, as shown in the peptide synthesis: unequivocal growth of the peptide, prevention of ancillary reactions and ease of separation.

In the search for improving multistage syntheses, an innovating and very promising approach has been published by Gelman and colleagues.[57,58] They have taken advantage of the chemical functionalization of materials and invented a 'one-pot' method which avoids the intermediate stages. Let us mention as an example the acid-catalyzed pinacol–pinacolone rearrangement, followed by the Knoevenagel condensation with malononitrile, which requires a basic medium (Figure 4.10).

Since the two catalysts (acid and base) are incompatible in the same medium, the synthesis in solution implies two distinct stages (a) and (b): the intermediate ketone must be isolated before the second stage can begin. In a heterogeneous phase, the two stages are replaced by two kinds

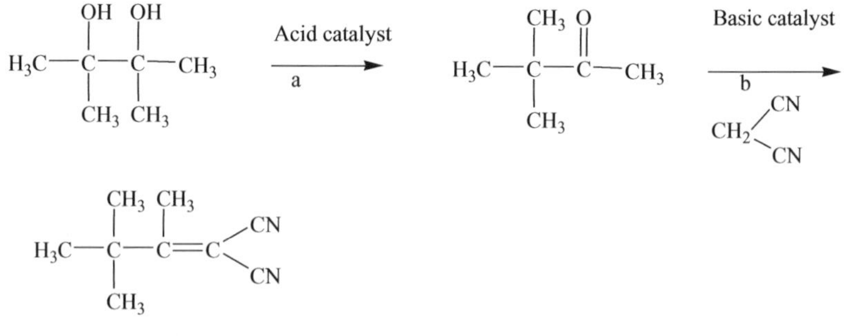

Figure 4.10 Pinacolic rearrangement in acidic medium followed by the Knoevenagel condensation in basic conditions. Reprinted with permission from *Angew. Chem., Int. Ed*, Acids and bases in one pot while avoiding their mutual destruction by Gelman, Faina; Blum, Jochanan; Avnir, David, *40*, 19, 3647–3649. Copyright (2001) Wiley-VCH

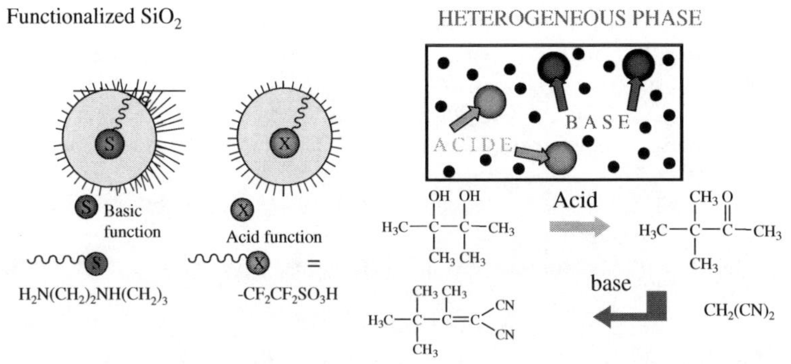

Figure 4.11 'One-pot' organic synthesis in a heterogeneous phase of a two-step process: the pinacolic rearrangement (acidic conditions) followed by the Knoevenagel condensation (basic conditions) (See Plate 5 for color representation)

of silica prepared separately, each having different functionalities. The first silica has an acid function (a perfluorosulfonic acid). On the second, a $H_2N(CH_2)_2NH(CH_2)_3-$ basic function has been grafted. It is then possible to mix in the same reactor these two solids in suspension in a liquid. The two chemical functions, being grafted on different solid particles, cannot interact and thus neutralize each other. The liquid plays an essential role: it contains the mixture of reagents and ensures the communication between the reagents and the supported catalysts (Figure 4.11).

The pinacolic rearrangement, producing the ketone, is only possible at the surface of particles containing the acid catalyst. The condensation of malononitrile can occur on account of particles having grafted basic functions, which generate the malononitrile carbanion, which reacts with the ketone.

Many syntheses can be improved using reaction cascades in a heterogeneous phase. Particles having suitable functionalities can be easily obtained by inorganic polymerization. This chemistry has made considerable progress in the last few years, in particular owing to the invention of nanoporous materials, which will be discussed in Chapter 6. This field deserves to be developed.

REFERENCES

[1] J.-P. Boilot, T. Gacoin, K. Lahlil, V. Buissette, D. Giaume, M. Matheron, *Actualité Chimique* **2005**, *290–291*, 77.

[2] D. Avnir, S. Braun, O. Lev, M. Ottolenghi, *Sol-Gel Optics, Processing and Applications*, L.C. Klein, (ed.), Kluwer Academic, Dordrecht, **1994**, 539.

[3] C. Sanchez, B. Lebeau, F. Chaput, J.-P. Boilot, *Adv. Mater.* **2003**, *15*, **1969**.
[4] J.-P. Boilot, F. Chaput, T. Gacoin, L. Malier, M. Canva, A. Brun, Y. Levy, J.-P. Galaup, *C. R. Acad. Sci. Paris* **1996**, *322*, 27.
[5] B. Lebeau, C. Sanchez, *Solid State Mater. Sci.* **1999**, *4*, 11.
[6] A. Huignard, V. Buissette, A.-C. Franville, T. Gacoin, J.-P. Boilot, *J. Phys. Chem. B* **2003**, *107*, 6754.
[7] A. Huignard, T. Gacoin, F. Chaput, J.-P. Boilot, P. Aschehoug, B. Viana, *Mater. Res. Soc. Symp Proc.* **2001**, *667*, G4. 5/1.
[8] M. Bruchez Jr, M. Moronne, P. Gin, S. Weiss, A.P. Alivisatos, *Science* **1998**, *281*, 2013.
[9] W.W.C. Chan, S. Nie, *Science* **1998**, *281*, 2016.
[10] M. Dahan, S. Lévi, C. Luccardini, P. Rostaing, B. Riveau, A. Triller, *Science* **2003**, *302*, 442.
[11] B. Dubertret, P. Skourides, D.J. Norris, V. Noireaux, A.H. Brivanlou, *Science* **2002**, *298*, 1759.
[12] B.O. Dabbousi, J. Rodriguez-Viejo, F.V. Mikulec, J.R. Heine, H. Mattoussi, R. Ober, K.F. Jensen, M.G. Bawendi, *J. Phys. Chem. B* **1997**, *101*, 9463.
[13] K. Riwotzki, M. Haase, *J. Phys. Chem. B* **1998**, *102*, 10129.
[14] A. Huignard, T. Gacoin, J.-P. Boilot, *Chem. Mater.* **2000**, *12*, **1090**.
[15] A. Huignard, V. Buissette, G. Laurent, T. Gacoin, J.-P. Boilot, *Chem. Mater.* **2002**, *14*, 2264.
[16] V. Buissette, M. Moreau, T. Gacoin, J.-P. Boilot, J.-Y. Chane-Ching, T. Le Mercier, *Chem. Mater.* **2004**, *16*, 3767.
[17] C.N.R. Rao, A.K. Cheetham, *J. Mater. Chem.* **2001**, *11*, 2887.
[18] M. Brust, C.J. Kiely, *Physicochem. Eng. Aspects* **2002**, *202*, 175.
[19] P. Mulvaney, *MRS Bull.* **2001**, *26*, **1009**.
[20] T. Takagahara, *Surf. Sci.* **1992**, *267*, 310.
[21] J.J. Shiang, J.R. Heath, C.P. Collier, R.J. Saykally, *J. Phys. Chem. B* **1998**, *102*, 3425.
[22] A. Taleb, V. Russier, A. Courty, M.P. Pileni, *Phys. Rev. B* **1999**, *59*, 13350.
[23] K.C. Beverly, J.F. Sampaio, J.R. Heath, *J. Phys. Chem. B* **2002**, *106*, 2131.
[24] M. Brust, D. Bethell, C.J. Kiely, D.J. Schiffrin, *Langmuir* **1998**, *14*, 5425.
[25] M.-P. Pileni, *J. Phys. Chem. B* **2001**, *105*, 3358.
[26] S. Besson, C. Ricolleau, T. Gacoin, C. Jacquiod, J.-P. Boilot, *J. Phys. Chem. B* **2000**, *104*, 12095.
[27] S. Besson, T. Gacoin, C. Ricolleau, C. Jacquiod, J.-P. Boilot, *J. Mater. Chem.* **2003**, *13*, 404.
[28] S. Besson, C. Ricolleau, T. Gacoin, C. Jacquiod, J.-P. Boilot, *Micropor. Mesopor. Mater.* **2003**, *60*, 43.
[29] D. Grosso, F. Babonneau, P.A. Albouy, H. Amenitsch, A.R. Balkenende, A. Brunet-Bruneau, J. Rivory, *Chem. Mater.* **2002**, *14*, 931.
[30] C. Boussard-Pledel, G. Fonteneau, J. Lucas, *J. Non-Cryst. Solids* **1995**, *188*, 147.
[31] M. Le Floch, C. Boussard-Pledel, G. Fonteneau, J. Lucas, J. Emery, J.-Y. Buzaret, *J. Phys.: Condens. Matter* **1998**, *10*, 5475.
[32] J. Shibin, Luo Tao, W. Jia-fu, *J. Non-Cryst. Solids* **2000**, *263*, 358.
[33] T. Kimmerlin, D. Seebach, *J. Peptide Res.* **2005**, *65*, 229.
[34] B. Nilnou, *Annu. Rev. Biophys. Biomol. Struct.* **2005**, *34*, 91.
[35] J.-M. Lehn, *Supramolecular Chemistry*, VCH, Weinheim, **1995**.

[36] F. Mathey, A. Sevin, *Introduction à la chimie moléculaire des éléments de transition,* Ellipses, Ecole Polytechnique, Paris, **1991**.
[37] C. Elschenbroich, A. Salzer, *Organometallics: a Concise Introduction*, VCH, Weinheim, **1989**.
[38] S. Brandes, F. Denat, G. Meyer, R. Guilard, *Actualité Chimique* **2005**, *290–291*, 108.
[39] M.L. Guillemot, C. Saout, R. Tardivel, C. Arnaud, A. Robin, *J. Eur. Hydrol.* **2002**, *33*, 11.
[40] R. Guilard, B. Roux-Fouillet, G. Lagrange, M. Meyer, A.T. Bucaille, F.R. 2804104, **2001**.
[41] F. Cuenot, M. Meyer, A. Bucaille, R. Guilard, *J. Mol. Liq.* **2005**, *118*, 89.
[42] F. Barbette, F. Rascalou, H. Chollet, J.-L. Babouhot, F. Denat, R. Guilard, *Anal. Chim. Acta* **2004**, *502*, 179.
[43] R. Guilard, H. Chollet, P. Guiberteau, P. Cocolios, WO 9611056, **1996**.
[44] H. Chollet, J.-L. Babouhot, F. Barbette, R. Guilard, WO 2001015806, **2001**.
[45] A. Favre-Réguillon, B. Dunjic, N. Dumont, M. Lemaire, *Tetrahedron* **1997**, *53*, 1343.
[46] A. Favre-Réguillon, B. Dunjic, M. Lemaire, R. Chomel, *Solvent Ext. Ion Exch.* **2001**, *19*, 181.
[47] A. Favre-Réguillon, B. Dunjic, N. Dumont, M. Lemaire, *J. Inclusion Phenom. Mol. Recognit. Chem.* **1998**, *32*, 477.
[48] A. Sorin, S. Pellet-Rostaing, A. Favre-Réguillon, M. Lemaire, *Sep. Sci. Technol.* **2004**, *39*, 2577.
[49] O. Vigneau, C. Pinel, M. Lemaire, *Chem. Lett.* **2002**, 202.
[50] E. Roth, *Pure Appl. Chem. 69*, **1997**, **1753**.
[51] M. Chemla, J. Périé, *La séparation des isotopes*, Presse Universitaire de France, Paris, **1974**.
[52] K. Nishizawa, Y. Maeda, F. Kawashiro, T. Fujii, T. Yamamoto, T. Hirata, *Sep. Sci. Technol.* **1998**, *33*, 2101.
[53] F.J. Feher, R. Terroba, R.-Z. Jin, S. Lucker, F. Nguyen, R. Brutchey, K.D. Wyndham, *Mater. Res. Soc. Symp. Proc.* **2001**, 628.
[54] F.J. Feher, R. Terroba, R.-Z. Jin, K.D. Wyndham, S. Lucke, R. Brutchey, F. Nguyen, *Polym. Mater. Sci. Eng.* **2000**, *82*, 301.
[55] G. Cerveau, R.J.P. Corriu, J. Dabosi, J.L. Aubagnac, R. Combarieu, Y. de Puydt, *J. Mater. Chem. 1998*, *8*, **1761**.
[56] G. Cerveau, R.J.P. Corriu, J. Dabosi, C. Fischmeister-Lepeytre, R. Combarieu, Rapid *Commun. Mass Spectrom.* **1999**, *13*, 2183.
[57] F. Gelman, J. Blum, D. Avnir, *New J. Chem.* **2003**, *27*, 205.
[58] F. Gelman J. Blum D. Avnir *Angew. Chem., Int. Ed.* **2001**, *40*, 19.

5

Nanostructured Materials

5.1 GENERAL REMARKS

Nanostructured materials exist as a *single phase* resulting from the incorporation of an organic entity into a silica matrix. In contrast to nanocomposites, *the nano-object cannot be separated from the matrix* in nanostructured materials, since they are linked together by several strong covalent bonds ($\sim$100 kcal mol^{-1}).

Nanostructured materials are obtained by synthesizing the matrix 'around' the nano-object. The principle consists of preparing molecular precursors substituted by chemical functions. These functions will permit the construction of the matrix by hydrolytic polycondensation (sol-gel process). In the case of silica (the only matrix investigated up to now, for reasons to be explained later), the precursors must have at least two Si $(OR)_3$ (R = Me, Et, iPr) groups attached to the nano-object by covalent bonds (Figure 5.1).

These covalent bonds must permit the construction of the SiO_2 matrix around the nano-object. In precursors containing only one $Si(OR)_3$ group (*cf.* Chapter 4) or in those having two $Si(OR)_3$ groups which are too dissymmetrically distributed on the molecular skeleton, polycondensation leads to a simple graft. As a matter of fact, hydrophilic and hydrophobic interactions will then regroup the organic parts on one side and the polar heads on the other side. This results in a graft on the surface, whereas the intended objective in nanostructured materials is to create an intimate organic-inorganic blend by regular inclusion of nano-objects within the matrix (Figure 5.2).

Molecular Chemistry of Sol-Gel Derived Nanomaterials Robert Corriu and Nguyên Trong Anh

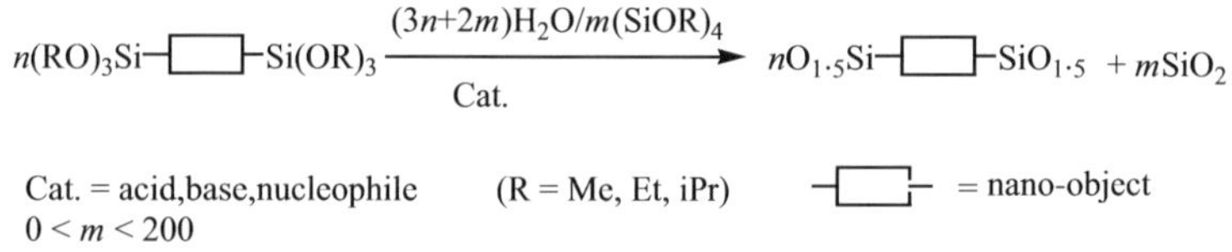

Figure 5.1 Synthesis of the silica matrix from a precursor substituted by at least two $Si(OR)_3$ groups

It is important to stress that this type of nanomaterial corresponds to a complex synthetic approach which can be generalized only if a certain number of criteria, which we shall now discuss, are satisfied.

5.2 SYNTHESIS OF HYBRID NANOMATERIALS

5.2.1 General Remarks

Before describing the results obtained so far, it is important to indicate some fundamental principles governing the preparation of an oxide matrix around a nano-object.[1,2] Since a single phase is needed, the problem to be solved is the synthesis of suitable precursors. Both the preparation of the organometallic entity (precursor) and that of the matrix to which the nano-object must remain bonded have to be considered. The success of the latter objective requires three conditions (Figure 5.1). The first corresponds to the introduction in the nano-object of groups capable of reacting with water to generate the oxide matrix by inorganic polymerization. The second

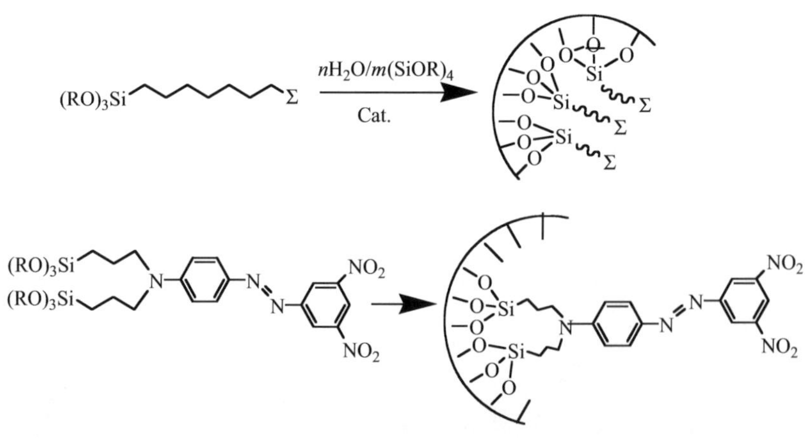

Figure 5.2 Polycondensation of monosubstituted or laterally substituted precursors, equivalent to a graft on silica

requires the reactive groups to be regularly distributed on the nano-object. The third is the *stability of the carbon-matrix* bonds which must be water-resistant. Indeed, it is this stability which determines the existence of the single matrix/nano-object phase. Clearly, this condition cannot be fulfilled by the majority of metals, since the metal–carbon bond is very labile. It is particularly sensitive to water in view of the M^{+}–C^{-} polarity. Water, necessary for the synthesis of the matrix, would cleave it, giving a metal alkoxide and a C–H bond [Equation (5.1)].

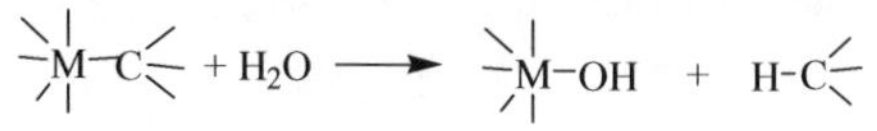

(M = Al, Ti, Zr, Ga, Cu...)

Equation 5.1

The only elements resisting this cleavage are those of groups IV (Si, Ge, Sn) and V (P, As, Sb). All the others (Al, Ti, V, Zr, Ga, In, transition metals, etc.) are more or less rapidly hydrolyzed. In practice, this short analysis can be summarized by saying that Si and P are the two elements capable of ensuring the stability of hybrid nanomaterials. Indeed, they permit the synthesis of the matrix while maintaining the (Si–C or P–C) covalent bonds with the organic molecule. Other elements can be used, e.g. B and Sn which give oxides by hydrolytic polycondensation. However, it must be noted that C–B and C–Sn bonds can be hydrolyzed under some experimental conditions. Some otherwise suitable elements are prohibited for obvious reasons: As and Sb for their toxicity and organogermanium compounds because of their cost.

Thus at present, Si is, *de facto*, the element most frequently employed, on the one hand because of the Si–C bond stability and of the numerous studies on SiO_2, and on the other hand because of its versatility and its excellent matrix properties. Note that phosphorus could be an excellent candidate: the chemistry of its oxides and nitrides is varied and well known. Furthermore, the P–C bond is chemically stable. However, until now, this type of combination has not been the subject of any notable studies. *Certainly, this deserves to be investigated.* However, the ability of phosphorous to permit extensive polycondensation is not comparable with that of silicon.

5.2.2 Why Silicon and Silica?

It is interesting to review the reasons justifying the choice of silicon as the key element for this type of materials. The main reason resides in the properties of SiO_2. Many silicon compounds can also easily lead to SiO_2 by hydrolytic polycondensation. However, a number of other aspects should be highlighted:[3–5]

1. The chemistry of organosilicon compounds has known such a development that a large range of combinations can now be prepared, from simple molecular compounds through a large variety of polymers to composite or porous solid materials. This chemistry is based on well established fundamental rules. Furthermore, a number of basic products are commercially available. They come from the silicones industry which, having numerous applications, offers a large variety of operational reactants.
2. Characterization of molecules, polymers and solids can rely on a series of well established analytical and spectroscopic methods (NMR, IR, MS, etc.).
3. The stability of Si–C bonds has been extensively explored. Si–C sp^3 bonds are very stable; Si–C sp^2 and Si–C sp bonds are more labile. However, only some very specific reagents permit their selective cleavage in well defined conditions involving basic or nucleophilic (Si–C≡C) or acidic (Si-aryl) catalysts.[4]
4. The formation of silica matrices, in particular by hydrolysis of R-SiX_3 (X = OMe, OEt, OiPr, H, Cl, OCOR, NR_2, SR, etc.) bonds, has also been widely investigated. This chemistry is at present well understood and commonly exploited.
5. The stability of hybrids depends on that of silica on the one hand, and on that of the organic entities on the other hand. Therefore, the thermal stability of hybrid materials lies below 250 °C. These materials resist acids quite well. However, it is known that basic treatments induce redistribution, even cleavage of the Si–O–Si bonds, possibly resulting in the material dissolving. Note that some bonds (e.g. SiC≡C) can also be sensitive to nucleophilic reagents (F^-).
6. Shaping the materials is very easy. It is always possible to control the texture of the material by varying the experimental conditions, as described earlier (*cf.* Section 3.6). Thus, *the level of hydrophilicity, the specific surface area and the solid porosity are adjustable*. This is very important for practical applications.
7. Materials are generally transparent, which facilitates their uses in optics.
8. Methods for silylation of organic compounds are quite numerous and most diverse.[3–5] The very large existing array will permit the synthesis of the majority, if not all, of the conceivable precursors. The synthesis may be more or less difficult, but it will be possible.

9. The most interesting property is the total compatibility of these matrices with all chemical entities: organic, inorganic, polymeric and even biological.
10. Recall, finally, that hydrolytic polycondensation on the silicon atom takes place *quantitatively* in mild conditions and with relatively simple experimental procedures.

5.2.3 Main Silylation Methods. Some Examples of Synthesis

5.2.3.1 General remarks

Synthesis of precursors requires grafting functional groups which permit the construction of the inorganic matrix around the chosen nano-object. Single phase hybrid materials can be obtained only when very stable bonds are present. Therefore the functional groups introduced on the nano-object are of the $C{-}SiX_3$ (X = Cl, OEt, OMe, OiPr, H, NR_2, OCOR, etc.) type.[6–10] In the majority of cases $Si(OR)_3$ alkoxides are used, since their reaction rate is adjustable and corresponds generally to a convenient timescale and experimental procedures. $Si(OEt)_3$ groups are the most frequently used. More rapid polymerizations are obtained with $Si(OMe)_3$. The $Si(OiPr)_3$ group is employed whenever it is desirable to slow down the hydrolysis reaction. It is used in particular to allow the chromatographic purification of precursors, since it is not hydrolyzed during the elution on the column. Afterwards, $Si(OiPr)_3$ can be transformed into $Si(OMe)_3$ or Si $(OEt)_3$ by a quantitative exchange reaction which is very easy to carry out [Equation (5.2)].

$$\text{R-Si(OiPr)}_3 \xrightarrow[\text{Cat.} = p\text{-toluenesulfonic acid or CsF}]{\text{EtOH (or MeOH)}} \text{R-Si(OEt)}_3 \text{ or R-Si(OMe)}_3$$

Equation 5.2

In the presence of catalysts, the SiH_3 group can be hydrolyzed in mild conditions, which might be advantageous [Equation (5.3)].

$$\text{R-SiH}_3 \xrightarrow[\text{Cat.}]{\text{H}_2\text{O}} \text{R-SiO}_{1.5} \quad [\text{Cat.} = \text{CsF, NH}_4\text{F, (Ph}_3\text{P)}_3\text{RhCl,...}]$$

Equation 5.3

The functional group Si–H is potentially interesting since it is not hydrolyzed in acidic media. Furthermore it can lead to different textures. Note, however, that SiH_3 is not easy to obtain and that H_2 is the hydrolysis product. Its handling requires, therefore, some precautions. Effectively, following redistribution reactions, this group can in some cases eliminate SiH_4, which is flammable and even explosive in air.

5.2.3.2 Methods permitting the introduction of $Si(OR)_3$ groups

After these general remarks, we shall now briefly review the methodologies which allow the introduction of a $Si(OR)_3$ group into an organic entity. This introduction is more difficult than that of a $SiMe_3$ group, since the SiOR bonds can react with a large number of nucleophilic reagents. In addition, their reactivity towards H_2O renders the separation and purification of precursors more difficult.

The most convenient route consists of using silylation reagents either commercially available or easily prepared, such as $I(CH_2)_3Si(OEt)_3$. Some of them, as well as two examples of application, are shown in Figure 5.3.

Hydrosilylation reactions[3,4,11] may be used. They require homogeneous or heterogeneous metal catalysts [Pd/C, Pt, $(Ph_3P)_3RhCl$, etc.]. Figure 5.4 illustrates the efficiency of this method for the synthesis of α,ω-bis-silylated aliphatic chains.

The Calas–Dunoguès reaction,[12] which uses Mg as a reagent, permits the introduction of the $Si(OR)_3$ group on aromatic systems in one step by a very convenient procedure (Figure 5.5).

Lithiation methods can also be used.[13–16] They permit the trialkoxysilylation by reaction of the organolithium reagent with $ClSi(OR)_3$ (where R = Me, Et, OiPr). The process occurs in two successive steps: lithiation followed by the reaction with $ClSi(OR)_3$ (Figure 5.6).

Note that this reaction permits also the preparation of functional systems which may undergo successive reactions, since it is possible to

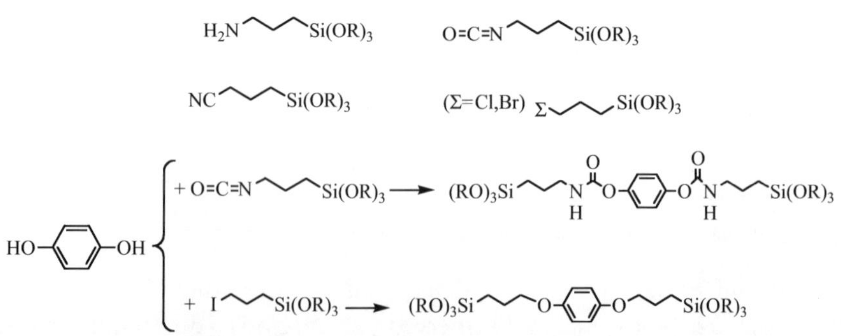

Figure 5.3 Commercially available molecules allowing the introduction of $Si(OR)_3$ groups on a nano-object: two examples of precursors

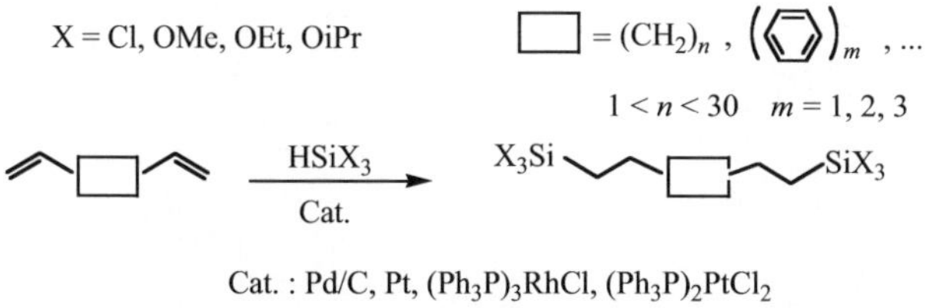

Figure 5.4 Hydrosilylation of double bonds

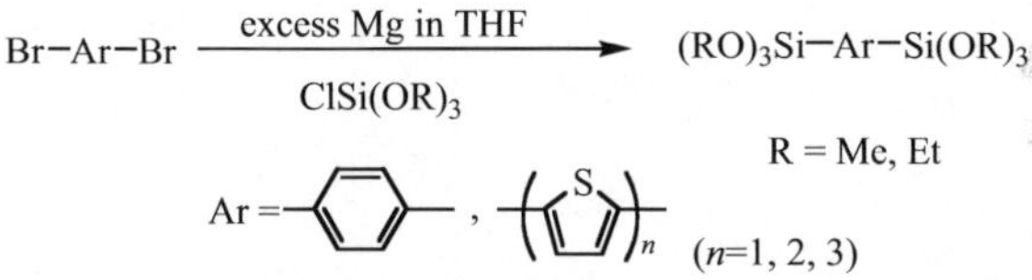

Figure 5.5 Synthesis of bifunctional aromatic compounds via the Calas–Dunoguès reaction. Reprinted with permission from *The Chemistry of Organosilicon Compounds*, vol. 3, Z. Rappoport and Y. Apeloig, Boury, B; Corriu, R.J.P., ch. 10, p. 565. Copyright (2001) John Wiley & Sons, Ltd

stop the reaction after monolithiation (if some precautions are taken). Thus, $Si(OR)_3$ groups can be attached to functionalized molecules (Figure 5.7).

Two Grignard reagents, $ClMgCH_2Si(OiPr)_3$ (**I**)[14] and *p*-$MgBrC_6H_4Si(OiPr)_3$ (**II**)[17] have been obtained by metallation of molecules bearing $Si(OiPr)_3$ groups. These reagents allow the introduction of groups containing the $Si(OR)_3$ function. Substitution of Si by (OiPr) groups markedly slows down the self-coupling reaction of the Grignard reagent. This method has permitted the production in one step of precursors difficult to obtain by the usual methods (Figure 5.8).

The aromatic Grignard reagent has likewise permitted the preparation of phosphines containing $Si(OiPr)_3$ groups. Nanomaterials containing

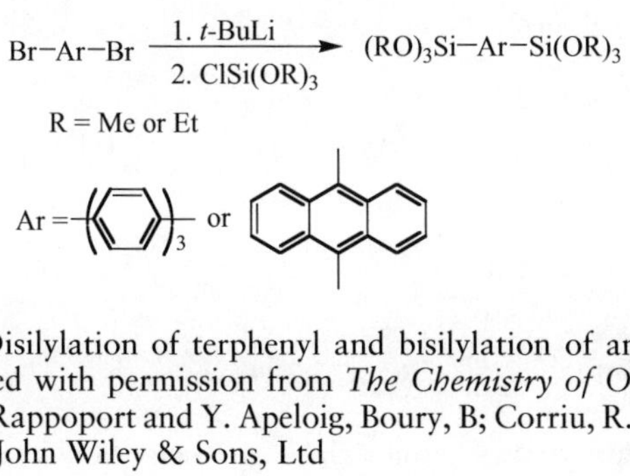

Figure 5.6 α,ω-Disilylation of terphenyl and bisilylation of anthracene on 9,10-positions. Reprinted with permission from *The Chemistry of Organosilicon Compounds*, vol. 3, Z. Rappoport and Y. Apeloig, Boury, B; Corriu, R.J.P., ch. 10, p. 565. Copyright (2001) John Wiley & Sons, Ltd

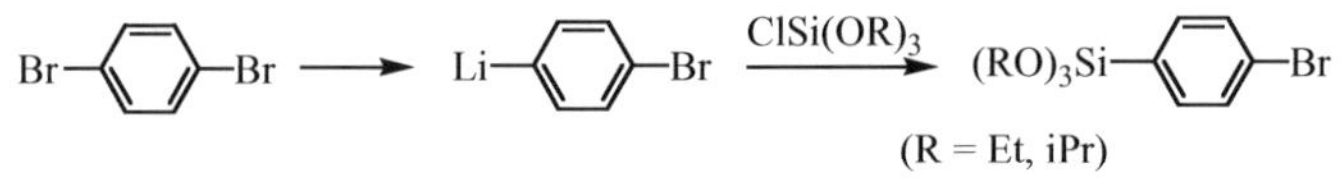

Figure 5.7 Introduction of $Si(OR)_3$ group to an aromatic compound by lithiation

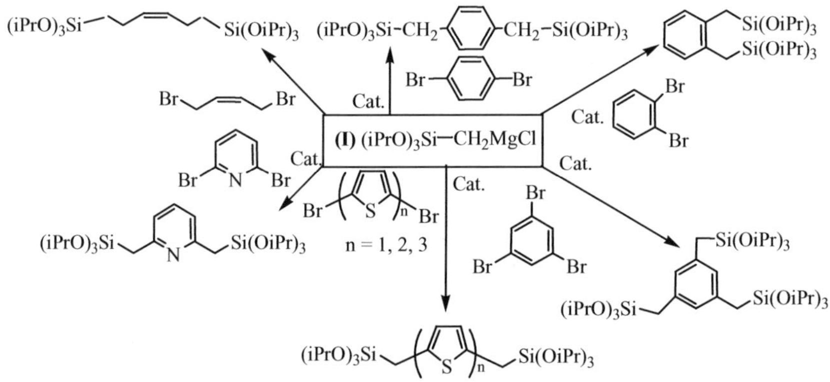

Figure 5.8 Some examples of precursors prepared using the Grignard reagent (**I**). Catalyst: $(PPh_3)_2NiCl_2$. Reprinted with permission from *J. Organometal. Chem.*, A new trialkoxysilylation reaction, the cross-coupling of (tri-isopropyloxysilyl)methyl Grignard reagent with organic halides by D. J. Brondani, R. J. P. Corriu, S. El Ayoubi, J. J. E. Moreau, M. Wong Chi Man, *451*, C1. Copyright (1993) Elsevier

transition metal (Ni, Pd, Pt) or lanthanide (Eu, Gd, Er) complexes, which shall be discussed later, have been obtained in this manner (*cf.* Chapters 5 and 6) (Figure 5.9).

The Benkeser reaction[18] is a less well known reaction which permits the silylation at the benzylic position in mild conditions with good yields (Figure 5.10).

The method known as the Heck reaction[19] represents one favored route for obtaining precursors containing vinylic or acetylenic groups. This reaction involves a Pd catalyst prepared *in situ* with a mixture of $Pd(OAc)_2$ and triorthotolylphosphine (TOP) in the presence of NEt_3. This reaction is

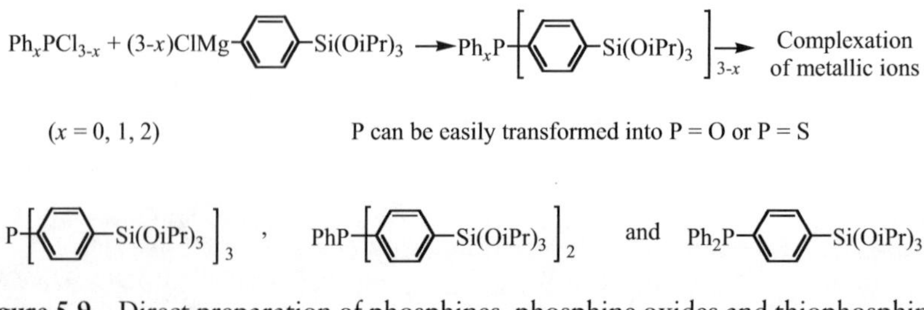

Figure 5.9 Direct preparation of phosphines, phosphine oxides and thiophosphines for complexation of metals

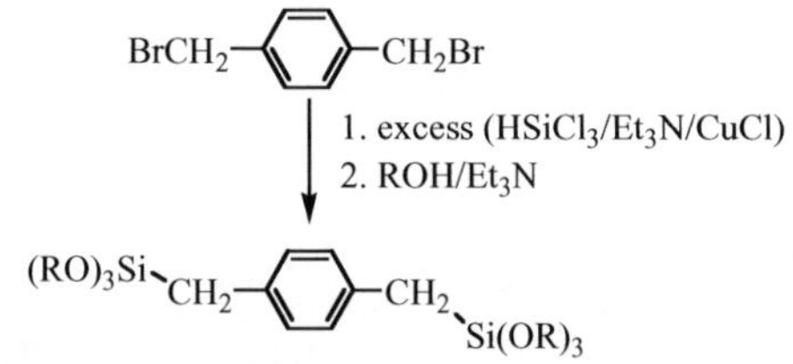

Figure 5.10 Benkeser silylation at the benzylic position. Reprinted with permission from *The Chemistry of Organosilicon Compounds*, vol. 3, Z. Rappoport and Y. Apeloig, Boury, B; Corriu, R.J.P., ch. 10, p. 565. Copyright (2001) John Wiley & Sons, Ltd

very flexible and has given rise to numerous interesting molecular geometries (Figure 5.11).

Dissymmetrical systems can also be obtained. Although not used until now, these precursors may lead to polyfunctional materials (Figure 5.12).

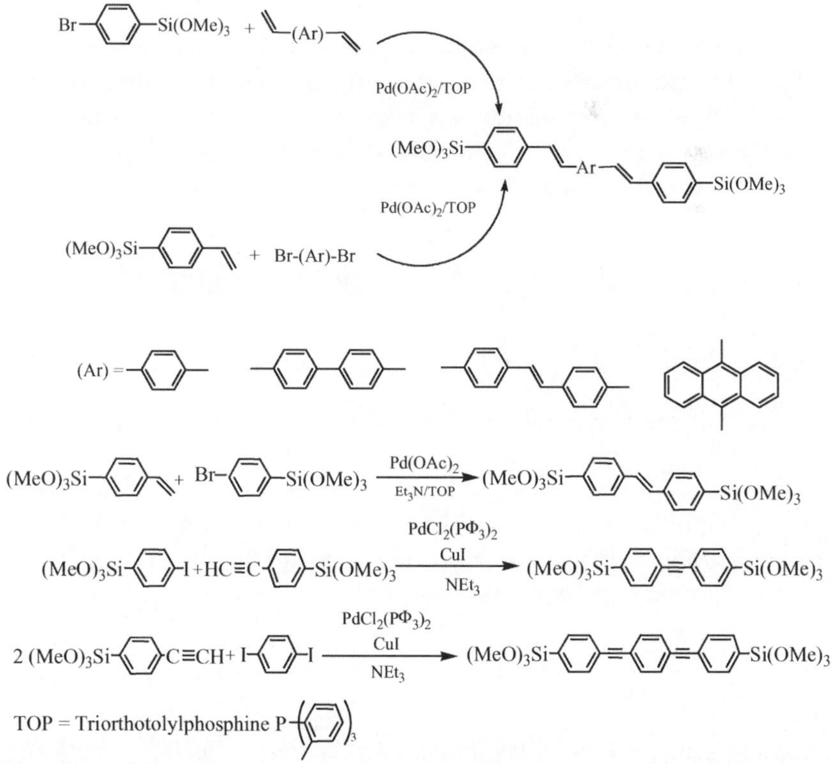

Figure 5.11 Examples of the Heck reaction for the synthesis of trialkoxysilylated precursors. Reprinted with permission from *The Chemistry of Organosilicon Compounds*, vol. 3, Z. Rappoport and Y. Apeloig, Boury, B; Corriu, R.J.P., ch. 10, p. 565. Copyright (2001) John Wiley & Sons, Ltd

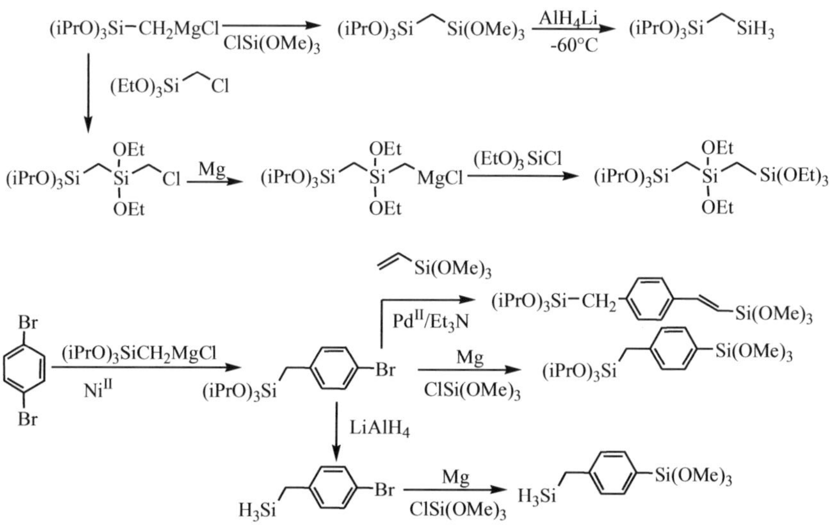

Figure 5.12 Some examples of dissymmetrical bisilylations on different positions

These few examples show the possibilities offered by existing synthetic methods. When all the possibilities are brought into play, they can offer access to very sophisticated materials. *In this field, methods of synthesis are no longer a limiting factor.*

5.3 NANOSTRUCTURED HYBRID MATERIALS

5.3.1 Examples of the Materials

Examples of the nanostructured hybrid materials already obtained are shown in Table 5.1.

These materials exist as a single phase in which the matrix and nano-objects cannot be separated. As shown before (*cf.* Section 3.5.3),[20] they are solids whose texture (specific surface area and porosity) is adjustable according to the experimental conditions.

5.3.2 Description of Nanostructured Hybrid Materials

The yield of hydrolytic polycondensation is always quantitative. However, the reaction shown in Figure 5.1 is purely theoretical, as it supposes that the materials obtained contain only totally polycondensed silicon atoms of the T^3 $Si(OSi)_3$ type and that Si–OR or Si–OH residual bonds

Table 5.1 Examples of organic and organometallic groups introduced into nanostructured hybrid materials

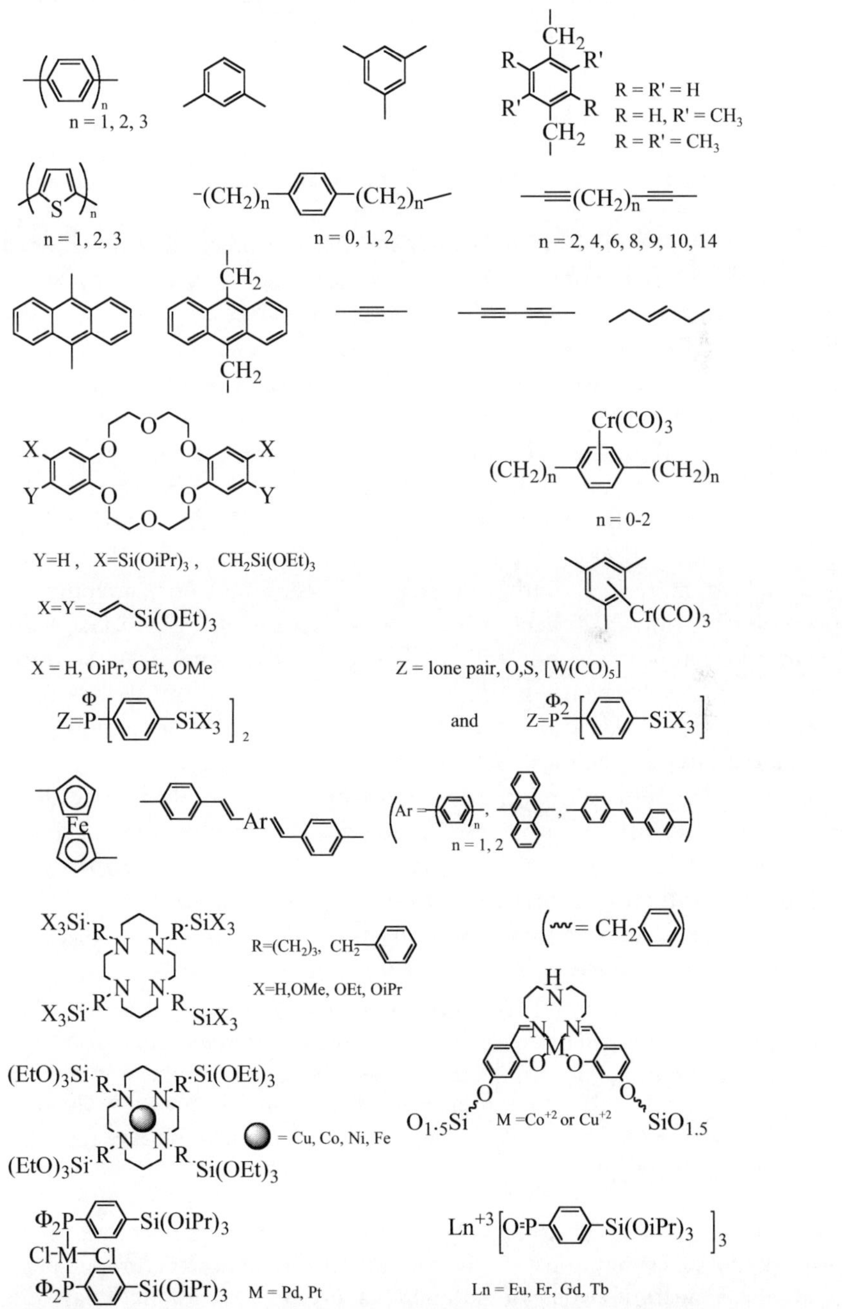

Reprinted with permission from *New J. Chem.*, Supramolecular self-organization in non-crystalline hybrid organic-inorganic nanomaterials induced by van der Waals interactions, Lerouge, Frédéric; Cerveau, Geneviève; Corriu, Robert J.P., *30*, 1364–1376. Copyright (2006) RSC

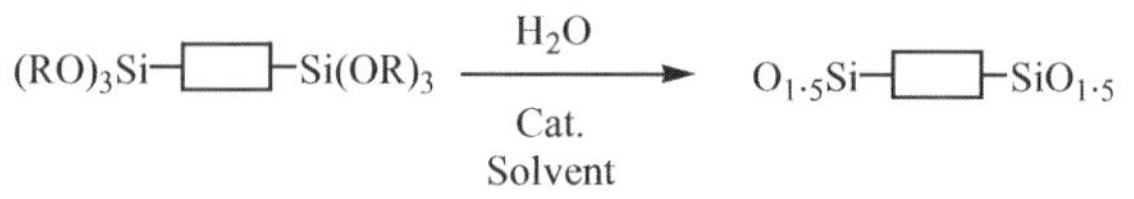

$T^3 = C\text{-}Si(OSi)_3$ $T^2 = C\text{-}Si(OSi)_2OR$ $T^1 = C\text{-}Si(OSi)(OR)_2$ $T^0 = C\text{-}Si(OR)_3$

Figure 5.13 The possibilities of polycondensation at silicon

are absent. In practice, this is not true. All the precursors are indeed incorporated into the polysiloxane network, but *the polycondensation of silicon is hardly ever complete.* The polycondensation to give a silica or a hybrid material is almost always incomplete. In the case of R-Si(OR)$_3$, a mixture of the T^1 Si(OSi)OR$_2$, T^2 Si(OSi)$_2$OR, T^3Si(OSi)$_3$ groups is obtained. The missing SiOSi bonds are in fact replaced by two SiOR (R = H or alkyl) groups. This incomplete polycondensation is the reason why yields higher than 100% are frequently observed (Figure 5.13).*

This has been shown unequivocally by ^{29}Si CPMAS NMR, which permits the detection of T^3, T^2, T^1 and T^0 units (Figure 3.22). In silica and hybrid materials, with only a few exceptions, only incomplete polycondensation has been observed. Generally, T^2·C-Si(OSi)$_2$(OR) groups are dominant, but the T^1·C-Si(OSi)(OR)$_2$ and T^3·C-Si(OSi)$_3$ groups are also present.[21,22] The presence of residual Si-OR has been confirmed by ^{13}C NMR and that of SiOH by IR spectroscopy. Finally (Si, C) microanalysis systematically indicates an excess of carbon. The polycondensation yield is quantitative, since it is impossible to detect or recover any quantity of unreacted precursor. In other words, *although the polycondensation reactions at silicon are not complete, all the precursor molecules are incorporated into the solid network.*

The characteristics of these solids are the following:

1. Considering the number of existing materials, it can be asserted that their synthesis can be generalized to all organic, organometallic entities of the main elements, of the transition elements and of the lanthanides.
2. The structure of the organic entity is not modified during the hydrolytic polycondensation leading to the material. ^{13}C NMR proves that this reaction takes place in conditions mild enough to preserve the organic structure.
3. To the best of our knowledge, until now, all polytrialkoxysilylated precursors submitted to hydrolysis have led to nanostructured materials. Transformations of voluminous precursors into materials are illustrative examples (Figure 5.14).[23–25]

*In the figures of this chapter, OR represents OR or OH.

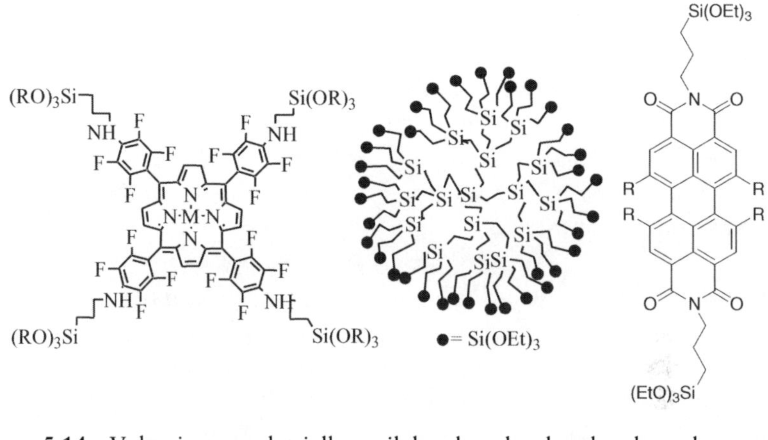

Figure 5.14 Voluminous polytrialkoxysilylated molecules that have been transformed into hybrid materials. Reprinted with permission from *The Chemistry of Organosilicon Compounds*, vol. 3, Z. Rappoport and Y. Apeloig, Boury, B; Corriu, R.J.P., ch. 10, p. 565. Copyright (2001) John Wiley & Sons, Ltd

However, cleavage of the Si$-$C$\equiv$C bond during the hydrolytic polycondensation in the presence of a F^- ion catalyst is an exception to this generalization. Note, however, that the polycondensation remains possible in acidic media (Figure 5.15).[26,27]

We can thus conclude that, but for this exception, all the molecular spacers studied have been incorporated without modification.

5.3.3 Some Characteristics

These materials are entirely new: they have only existed for a dozen years. It is pertinent to give now their principal characteristics.

A priori, these solids are not, in any way (specific surface area, porosity, etc.), different from other materials obtained by inorganic polymerization. However, they are structurally very different on account of the organic entities bonded to the matrix.

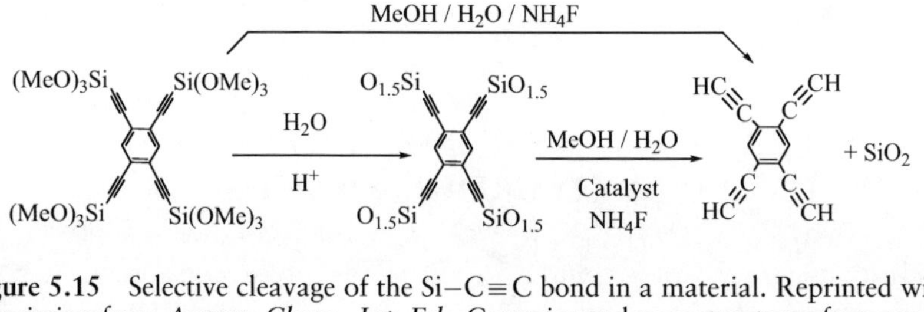

Figure 5.15 Selective cleavage of the Si$-$C$\equiv$C bond in a material. Reprinted with permission from *Angew. Chem., Int. Ed.*, Ceramics and nanostructures from molecular precursors by Robert J.P. Corriu, *39*, 1376–1398. Copyright (2000) Wiley-VCH

As we shall see, their properties depend essentially on the incorporated organic units. It can be asserted that these components retain physical and chemical properties very similar to those of the free molecular entity. The slight differences are due to the fact that the solid phase brings about a very poor mobility and also a possible organization of these entities (see below). We shall return to this point, but the behavior of these materials can be summarized as:

1. *It is possible to carry out chemical reactions on the organic entities chemically bonded to the material.* They are easily accessible to the majority of reagents: thus, additions on multiple bonds, oxidations and reductions can be performed. Some polymerization reactions have been obtained within the matrix.[28–30] Complexation of metal ions by chelating units[31,32] have particularly attracted attention. Indeed these complexes possess potential physical properties and are therefore possible nanomaterials which deserve to be investigated.
2. Moreover, as we shall see, *the existence of organization between the organic units incorporated in the matrix modifies the chemical and physical properties.*

5.4 KINETIC CONTROL OF THE TEXTURE OF NANOSTRUCTURED HYBRID MATERIALS

The kinetic control of hydrolytic polymerization reactions (sol-gel process) has been presented in detail (*cf.* Section 3.5.3).[20] Also, nanostructured materials, discussed here, have been taken as an example to demonstrate the possibility of using kinetic control to optimize the texture of materials.[33–39] As each kinetic parameter can influence the physical characteristics of the final material, *it should be possible to modify the specific surface area, the porosity and the pore size distribution by varying the experimental conditions; the stability of the final product is ensured by a hydrothermal treatment.*

5.5 SUPRAMOLECULAR SELF-ORGANIZATION INDUCED BY HYDROGEN BONDS

Hybrid materials corresponding to a single phase have been termed 'nanostructured' because the organic entities bonded to the matrix can

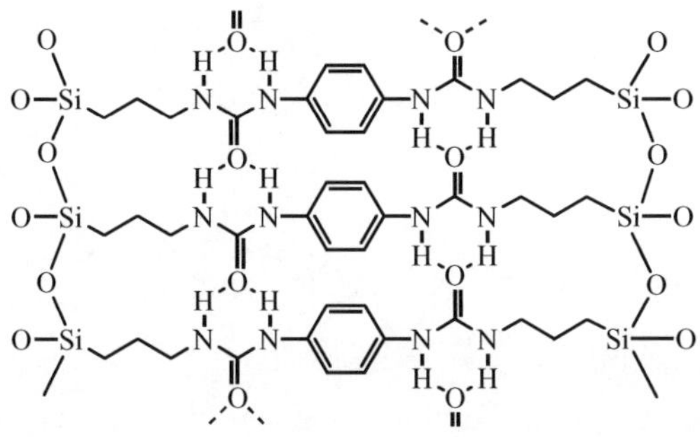

Figure 5.16 Network of hydrogen bonds between urea groups. Reprinted with permission from *Angew. Chem., Int. Ed.*, A better understanding of the self-structuration of bridged silsesquioxanes, J.J.E. Moreau, B.P. Pichon, M. Wong Chi Man, C. Bied, H. Pritzkow, J.-L. Bantignies, P. Dieudonne, J.-L. Sauvajol, *43*, 203. Copyright (2004) Wiley-VCH

induce organization within the material. According to our current knowledge, there are two self-organization modes for the organic entities embedded in the material: self-organization induced by van der Waals interactions between the organic spacers; and a completely different self-organization due to hydrogen bond networks.

This latter mode will be discussed first. Hydrogen bonds are introduced during the synthesis by attaching 'urea type' groups to the skeleton of the precursor using the reaction of iso-cyanatopropyltriethoxysilane with amine substrates.[40–46]

Polycondensation of $Si(OEt)_3$ groups leads to solids having an organization induced by the formation of networks of hydrogen bonds between urea groups (Figure 5.16). The energy of formation of a hydrogen bond is $\sim$5–10 kcal mol^{-1}. This bond is particularly strong in the case of urea, due to the very favorable geometry between >C=O and H–N< groups.

Three types of organization have been obtained. Aliphatic systems, such as $(CH_2)_{12}$, lead to lamellar type solids.[40–46] These structures have been confirmed by X-ray diffraction, TEM[†] and by the formation of columns tens of micrometers in size.

For systems having a rigid core, the organization comes from a crystalline system (Figure 5.17).[44,45] The X-ray structure shows clearly the alternation of hydrogen bonds which are responsible for the crystal's cohesion.

[†]Transmission electron microscopy. This technique can give images at the nanometer scale.

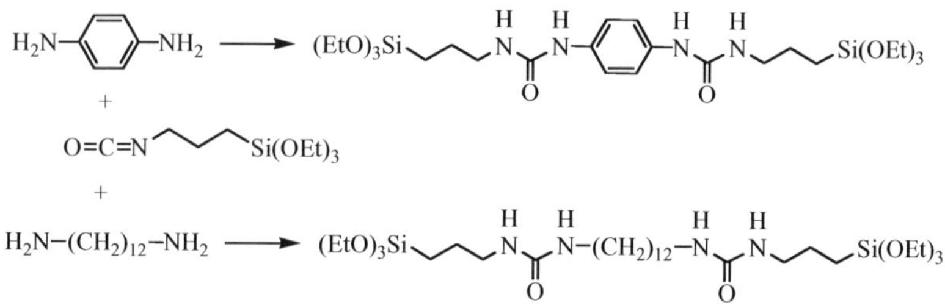

Figure 5.17 Organization of the aliphatic spacers by hydrogen bonding. Reprinted with permission from *J. Am. Chem. Soc.*, Self-organized hybrid silica with long-range ordered lamellar structure by J. J. E. Moreau, l. Vellutini, M. Wong Chi Man, C. Bied, J. L. Sauvajol, *123*, *7957*. Copyright (2001) ACS

Finally, in the case of chiral systems and optically active compounds derived from diamines, chiral columns of left- and right-handed helicoidal fibers have been obtained, depending on the (*RR*) or (*SS*) configuration of the two carbons bearing the NH_2 groups.[41–43] The length of these columns is in the order of 15 μm and their thickness varies from 0.3 to 1 μm (Figure 5.18).

This methodology can produce materials presenting an excellent organization. This type of assemblage may lead to the control of morphologies over long distances.

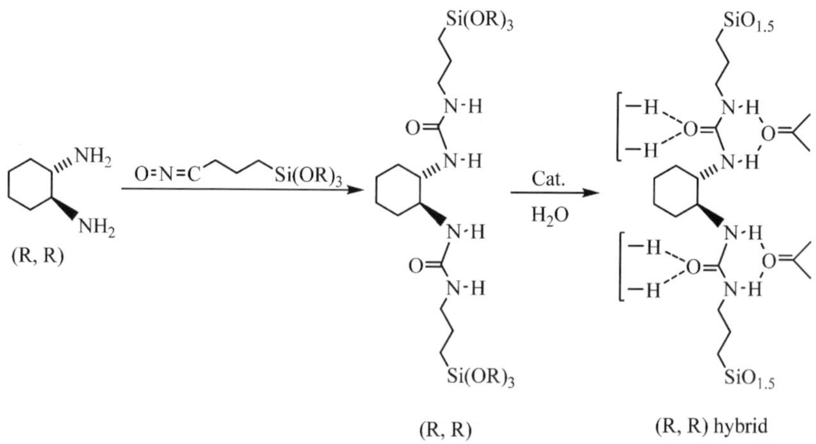

Figure 5.18 Left- and right-handed organization in chiral columnar nanomaterials. Reprinted with permission from *J. Am. Chem. Soc.*, New hybrid organic-inorganic solids with helical morphology via H-bond mediated sol-gel hydrolysis of silyl derivatives of chral (R,R)- or (S,S)-diureidocyclohexane by J. J. E. Moreau, L. Vellutini, M. Wong Chi Man, C. Bied, *123*, 1509. Copyright (2001) ACS

5.6 SUPRAMOLECULAR SELF-ORGANIZATION INDUCED BY WEAK VAN DER WAALS TYPE BONDS

5.6.1 What do We Mean by Self-Organization?

Before describing the results obtained, let us clarify the 'self-organization' concept. Self-organization was revealed by the 'unexpected' chemical behavior of some hybrid materials in which no interactions exist between organic entities other than van der Waals interactions. In fact, their unusual chemical behaviors have instigated research on this subject (*cf.* Section 5.6.2).

A priori, the usual identification criteria for hybrid materials cannot indicate the manner in which the organic units arrange themselves in the matrix. The texture (specific surface area, porosity, etc.) corresponds to external properties of the material (macroscopic morphology). It therefore cannot give any information on the relative layout of the organic entities within the solid. Moreover, from weight analysis data and from (^{13}C, ^{29}Si) NMR and IR spectra, it is impossible to get information about the way the organic units are arranged with respect to each other. The experimental data can agree with a hybrid solid in which the organic entities can adopt all the possible orientations as well as with structures in which the organic units are clearly organized (Figure 5.19).

5.6.2 Chemical Behavior and Self-Organization

The existence of a self-organization of organic units included in hybrid materials has been detected by studies of their reactivity. Two examples will be discussed. Hydrolytic polycondensation of precursors containing bis- and terthiophene units gave solids in which thiophene units bonded to the matrix are unambiguously identified. However, potentiometric study of the conductivity of thin layers of these materials deposited on an electrode has shown a clear evolution of the material's conductivity with time. This change arises from a polymerization reaction induced in the solid by oxidation (Figure 5.20).[28,29]

The presence in the solid of polymeric entities formed by polycondensation of thiophene units is characterized unambiguously by Raman resonance spectroscopy. The solid obtained corresponds to a polythiophene linked to the SiO_2 matrix and *cannot be separated from it.* This means that it is attached by covalent bonds. This surprising result suggests

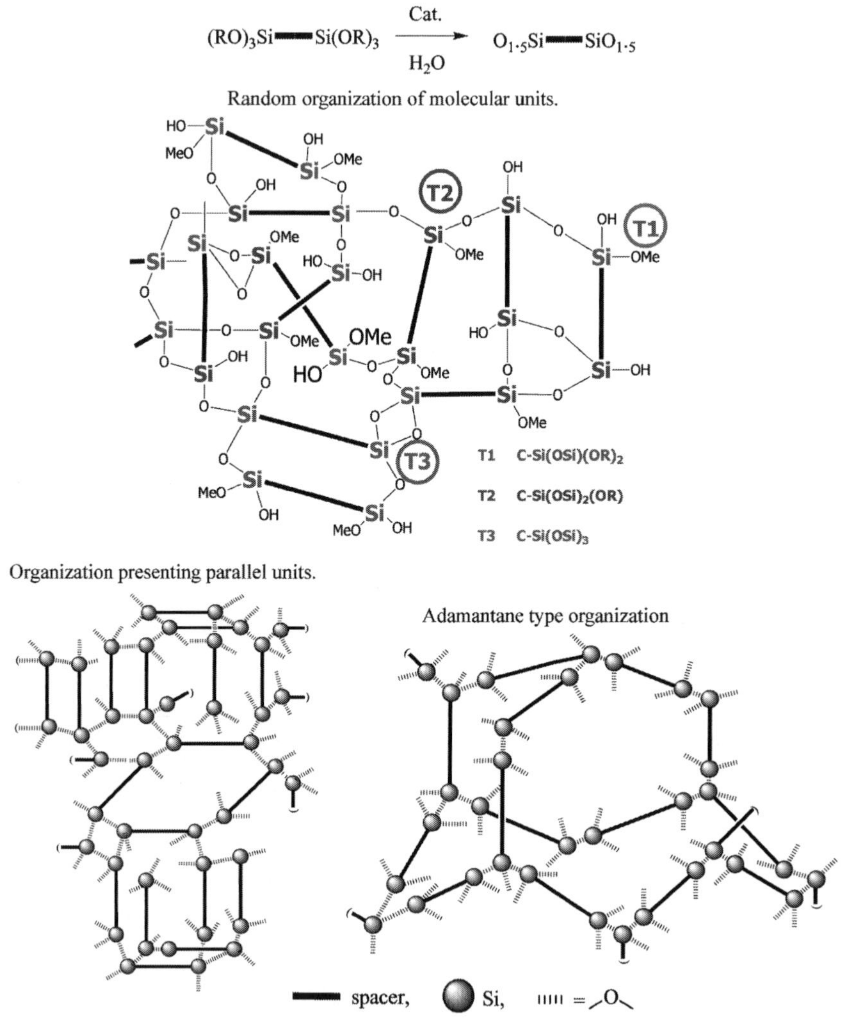

Figure 5.19 Three different types of arrangement are given as possible examples of organization of the organic units in the material. The organizations are different but all the data (weight analysis, BET, porosity, IR, NMR, etc.) are the same. Reprinted with permission from *New J. Chem.*, Supramolecular self-organization in non-crystalline hybrid organic-inorganic nanomaterials induced by van der Waals interactions, Lerouge, Frédéric; Cerveau, Geneviève; Corriu, Robert J.P., *30*, 1364–1376. Copyright (2006) RSC (See Plate 6 for color representation)

the existence of an organization of the thiophene units within the solid, since the electrochemical coupling of two thiophene radicals is known to require very strict stereoelectronic conditions (Figure 5.21). These conditions cannot be satisfied without a prior self-organization of the two

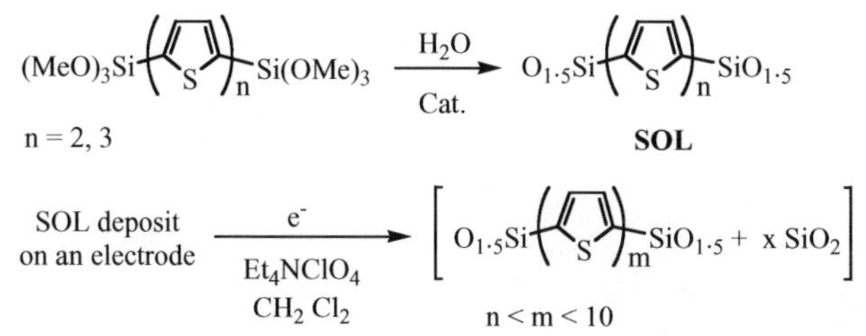

Figure 5.20 Polymerization in the solid observed with disilylated bis- and ter-thiophene in potentiometric experiments on an electrode. Reprinted with permission from *Angew. Chem., Int. Ed.*, Ceramics and nanostructures from molecular precursors by Robert J.P. Corriu, *39*, 1376–1398. Copyright (2000) Wiley-VCH

units in the solid. Indeed, these units must be localized in parallel planes to allow the formation of the C–C bond by the coupling of two radical-cations (Figure 5.21). This coupling is followed by the cleavage of the Si–C bonds by nucleophilic attack of H_2O on each Si atom. This reaction permits the formation of α,ω-disilylated polythiophene, the polymer formed remaining linked to SiO_2 at each of its ends.

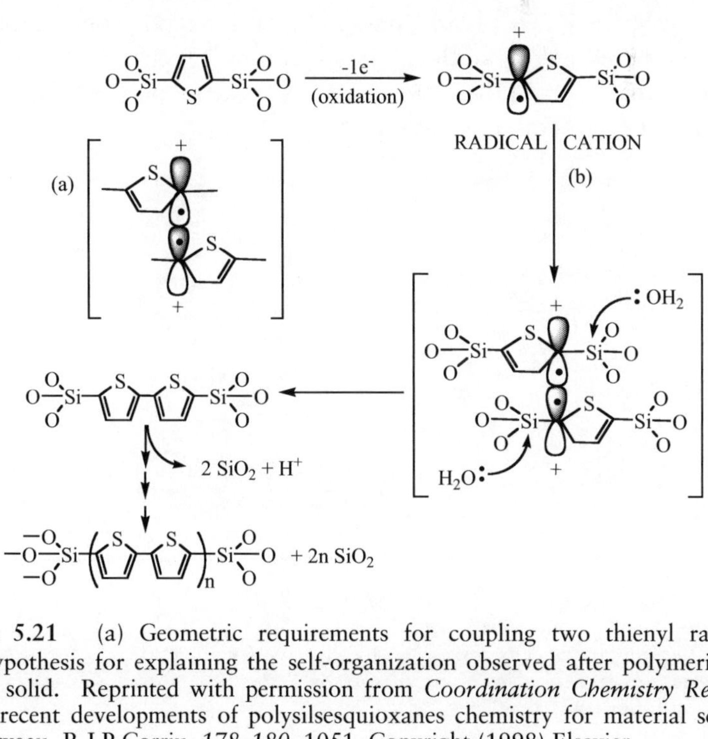

Figure 5.21 (a) Geometric requirements for coupling two thienyl radicals. (b) Hypothesis for explaining the self-organization observed after polymerization in the solid. Reprinted with permission from *Coordination Chemistry Reviews*, Some recent developments of polysilsesquioxanes chemistry for material science, G. Cerveau, R.J.P Corriu, *178–180*, 1051. Copyright (1998) Elsevier

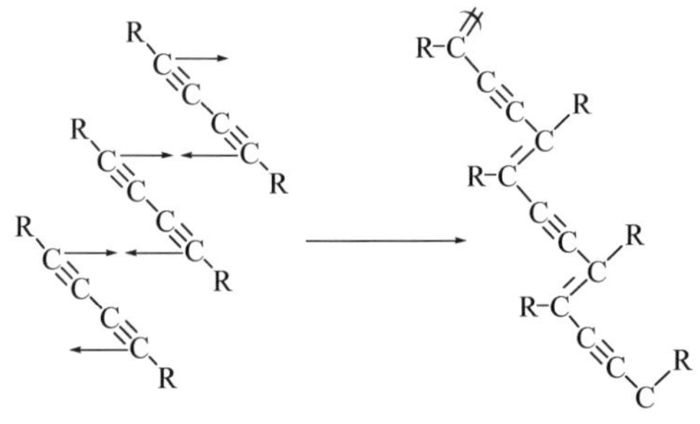

R = Phenyl, Alkyl, $Si(OEt)_3$, $Si(OMe)_3$

Figure 5.22 Topochemical polymerization observed during the coupling of bisacetylenic units leading to enynes. This transformation needs a good parallelism between the bisacetylenic units. Reprinted with permission from *Angew. Chem. Int. Ed.*, Ceramics and nanostructures from molecular precursors by Robert J.P. Corriu, *39*, 1376–1398. Copyright (2000) Wiley-VCH

The second example concerns the bisethynyl units which are transformed into enynes by a 1,4-addition process (Figure 5.22).[30]

Bisacetylenic molecules containing $(RO)_3Si$ groups at each end have been synthesized. They can undergo hydrolytic polycondensation at silicon and crosslinking of the bisethynyl units.[30] Figure 5.23 shows two different treatments of these molecules: in one case, polycondensation at silicon (sol-gel) takes place first, followed by the crosslinking of the

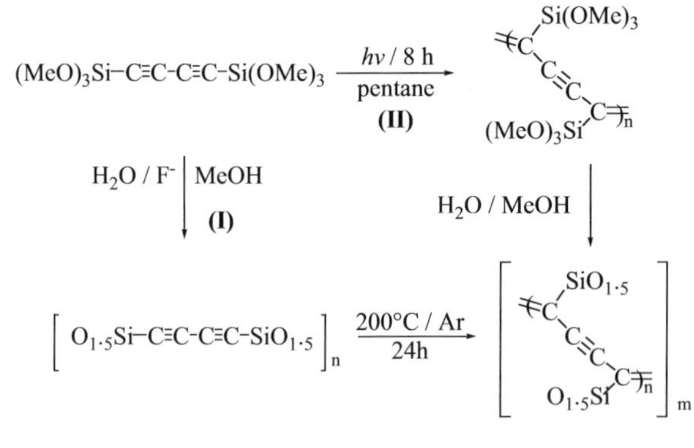

Figure 5.23 Two possible routes for the material: (**I**) hydrolytic polycondensation at silicon followed by condensation of the C≡C–C≡C units to give enynes; (**II**) reversed order. Reprinted with permission from *Angew. Chem. Int. Ed.*, Ceramics and nanostructures from molecular precursors by Robert J.P. Corriu, *39*, 1376–1398. Copyright (2000) Wiley-VCH

bisethynyl units (pathway I). In the other case, the two steps are reversed, so the crosslinking stage precedes the polycondensation (pathway II). The results are noteworthy, because the two solids have the same spectral characteristics. All the –C≡C–C≡C– units have been transformed into =C–C≡C–C= enyne groups. The two materials are doubly crosslinked at silicon with a majority of T^2units and no T^0units. Given that crosslinking of the –C≡C–C≡C– entities implies their parallelism, it can be concluded that polycondensation at silicon occurs on an *organized* network of C_4 entities.

The third example concerns the comparison of two materials obtained from two precursors (**A**) and (**B**) having similar geometries, but considerably different behaviors (Figure 5.24).[47,48]

The hydrophilicity, corresponding to the weight increase of the material in water atmosphere saturated at 60%, is high in the case of (**A**) (~25%). Remember that the hydrophilicity of silica varies from 15 to 40%, depending on the texture. In contrast, (**B**) is hydrophobic since, in the

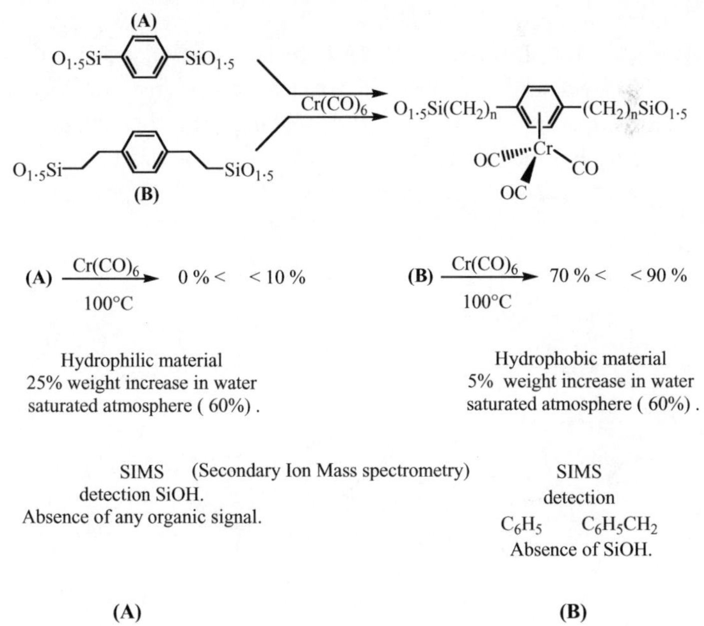

Figure 5.24 Comparison of the properties of two similar hybrid materials (**A**) and (**B**). Reprinted with permission from *J. Organometal. Chem.*, Nanoporous materials: a good opportunity for nanosciences by Robert J.P. Corriu, A. Mehdi, C. Reye, *689*, 4437. Copyright (2004) Elsevier

same conditions, the weight increase is <5%. The reaction of (**A**) with Cr $(CO)_6$ to give the [arene/$Cr(CO)_3$] adduct is difficult; the yields are always low (<10%). On the contrary, (**B**) reacts readily and the yields can reach 90%. Finally, the SIMS (secondary ion mass spectrometry),[49,50] classically used for the detection of entities located at the surface of solids, reveals in the case of (**A**) only SiOH groups. No fragment corresponding to an organic entity could be detected. This agrees with the hydrophilicity of (**A**). In contrast, the hydrophobic (**B**) shows no SiOH. Only the aromatic fragments C_6H_5 and $C_6H_5CH_2$, undetectable in the case of (**A**), have been detected. Clearly, despite the fact that their (NMR, IR, etc.) characteristics are almost identical, the two precursors (**A**) and (**B**) lead to solids having totally different organizations (Figure 5.25).

The experimental data can be rationalized by assuming two different structures. The organization of precursor (**A**) must permit the detection of SiOH, but not phenyl groups, by SIMS. An organization in which C_6H_4 groups are not located at the material surface but are perpendicular to it can explain the hydrophilicity, the absence of an aromatic peak in the SIMS spectrum and also the lower reactivity of (**A**) towards $Cr(CO)_6$ (steric hindrance between C_6H_4 units).

On the other hand, to explain the behavior of (**B**), a structure corresponding to a crosslinked network of linear systems, in which some aromatic groups are located on the surface, can be suggested. This explains the hydrophobicity, the evidence of C_6H_5 and C_7H_7 fragments by SIMS and the excellent reactivity towards $Cr(CO)_6$ (the linear chains are spaced out).

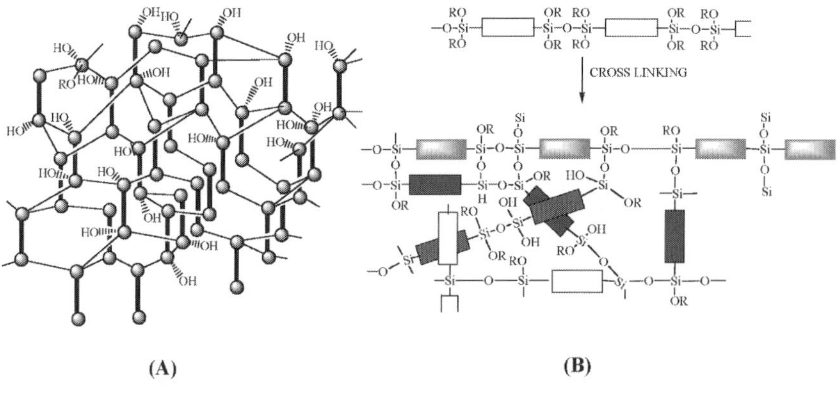

Figure 5.25 Schematic view of the organization of organic units inside the solids (**A**) and (**B**). Reprinted with permission from *Organosilicon Chemistry V, From Molecules to Materials* by B. Boury and R. Corriu, p. 389. Copyright (2003) Wiley-VCH

5.6.3 Study of Self-Organization

5.6.3.1 General remarks

The self-organization of organic units deserves to be investigated, since it depends only on the van der Waals type hydrophobic (or lipophilic) interactions occurring between organic units during the hydrolytic polycondensation. In the study undertaken, this reaction has been carried out *in solution, with the stoichiometric amount of* H_2O *required for the polycondensation*. The other parameters (nature of the solvent, concentration of the reagents, nature and concentration of the catalyst, temperature, etc.) are adjustable. It soon appears that the self-organization is, as for the texture, a kinetically controlled phenomenon, strongly dependent on experimental conditions. In fact, any experimental parameter (temperature, nature and concentration of the reagents, catalyst, solvent, etc.) can modify some characteristics of the material.

The study of this phenomenon was based on two types of techniques: X-ray diffraction (SAXS)[‡] and birefringence measurements. X-ray diffraction detects *organization at the nanometric scale*. The birefringence phenomenon under cross-polarized light permits the study of *organization at the micrometric scale*.

The most reasonable working hypothesis assumes that this organization results from lipophilic (or hydrophobic) van der Waals type interactions between organic units in the presence of hydrophilic entities. As a matter of fact, in the experimental conditions of the hydrolytic polycondensation (presence of a stoichiometric amount of H_2O, of an ionic catalyst and an organic solvent), the van der Waals hydrophobic interactions induce the self-assembling of organic moieties when the polycondensation at silicon generates hydrophilic SiOH groups. One can, therefore, reasonably consider that hydrophobic van der Waals type interactions play an important role in the self-organization of the material during the entire polycondensation process.

Let us examine first the experimental facts supporting this hypothesis.

5.6.3.2 Self-organization in linear systems

The first studies were carried out on linear and rigid molecular spacers. These precursors present only two parameters: rigid directionality and the

[‡]Small angle X-ray scattering.

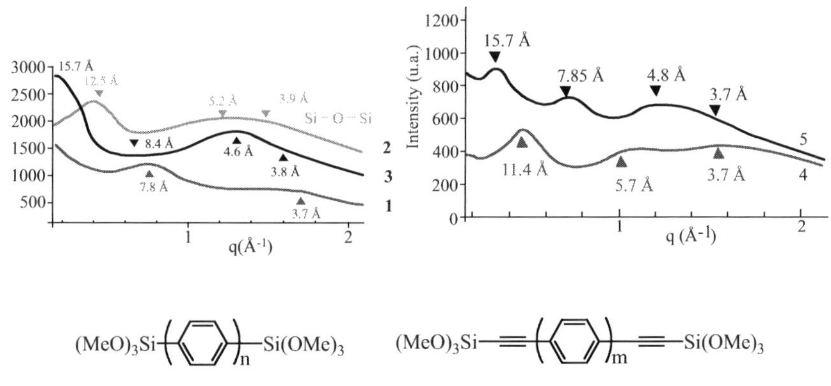

1 : n = 1 **2** : n = 2 **3** : n = 3 **4** : m = 1 **5** : m = 2

Figure 5.26 SAXS diffraction patterns of hybrid materials obtained after hydrolytic polycondensation of linear precursors (**1–5**). Reprinted with permission from *New J. Chem.*, Supramolecular self-organization in non-crystalline hybrid organic-inorganic nanomaterials induced by van der Waals interactions, Lerouge, Frédéric; Cerveau, Geneviève; Corriu, Robert J.P., *30*, 1364–1376. Copyright (2006) RSC

size of the spacer.[51–53] Precursors **1–5** have been synthesized and polycondensed in the same experimental conditions (Figure 5.26).

The X-ray diffraction patterns of the hybrid materials obtained have no sharp Bragg signals. However, the diffraction curve indicates an organization at the nanometric scale. Clearly, this organization exists between the organic units, since in the same conditions SiO_2 always presents the characteristics of an amorphous network (diffraction curve at ~3.7 Å, assigned to a network of Si–O–Si bonds).

It is important to stress the physical meaning of this type of diffraction curve: the presence of broad diffraction signals shows unambiguously that the material under study is an amorphous solid, or more precisely a noncrystalline solid *presenting a certain organization at the nanometric scale, since electronic density differences are revealed by X-ray diffraction.* Thus, in the first approximation, and assuming that Bragg's law applies to these diffraction diagrams, an *approximate evaluation* of the distance at which the organization may exist can be obtained. In the present cases, when the distance between the electronic densities of the two couples of $Si{-}O_{1.5}$ bonds is considered,[51–53] an obvious correlation appears between the value estimated from the diffraction spectrum on the one hand, and the length of the spacer calculated by molecular modeling methods on the other hand (Table 5.2).

Hence, the existence of broad diffraction signals clearly corresponds to a material presenting an organization at the nanometric scale.[54,55] Note also that the intensity of the diffraction signals varies with the experimen-

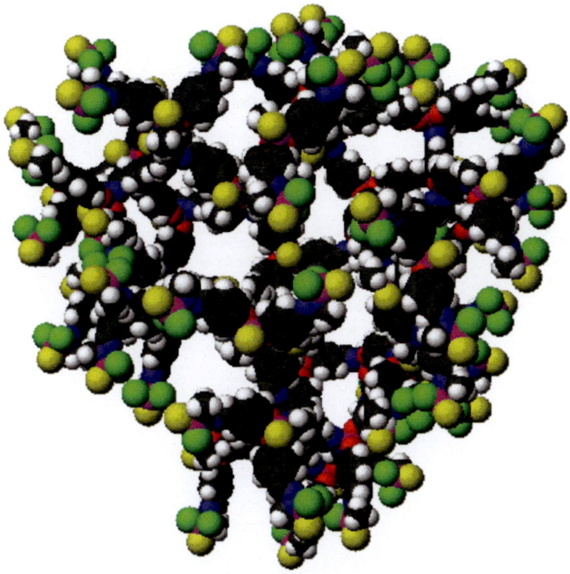

Plate 1 Molecular model of a phosphorus dendrimer presenting 48 $P(S)Cl_2$ groups on its surface

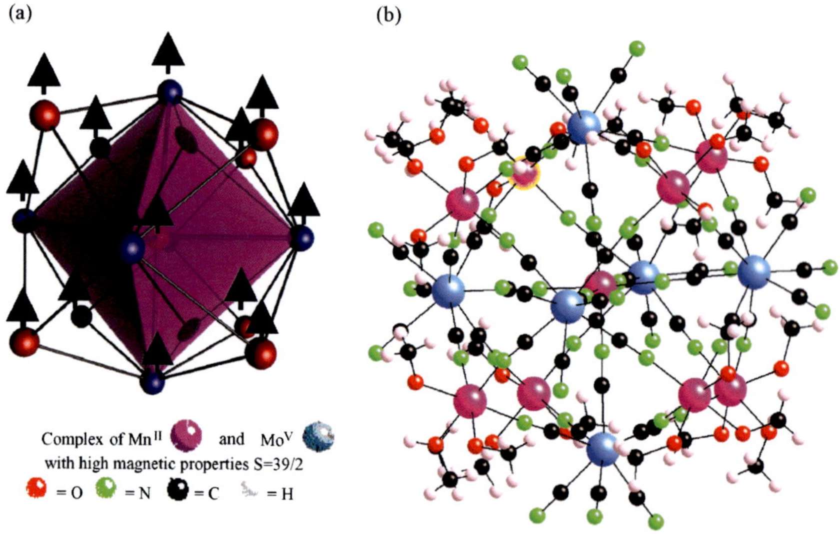

Plate 2 A bimetallic molecular complex having an important magnetic moment. The Mn^{II} and Mo^{V} ions are linked by $C\equiv N^{-}$ bridges. (a) Global scheme and (b) molecular structure. Reproduced by permission of *L'actualité chimique*

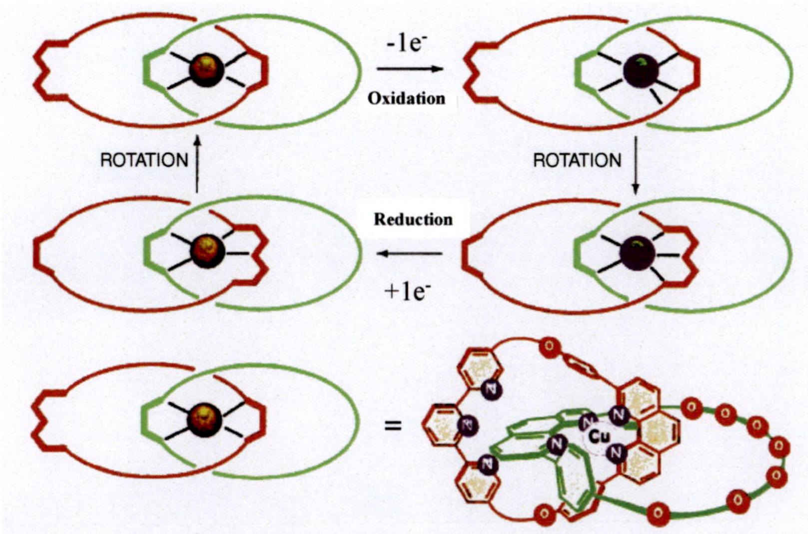

J-P. SAUVAGE and Col.. – UNIVERSITY Louis PASTEUR - STRASBOURG

Plate 3 Molecular engine with a rotor capable of rotating inside a stator. Reproduced by permission of *L'actualité chimique*

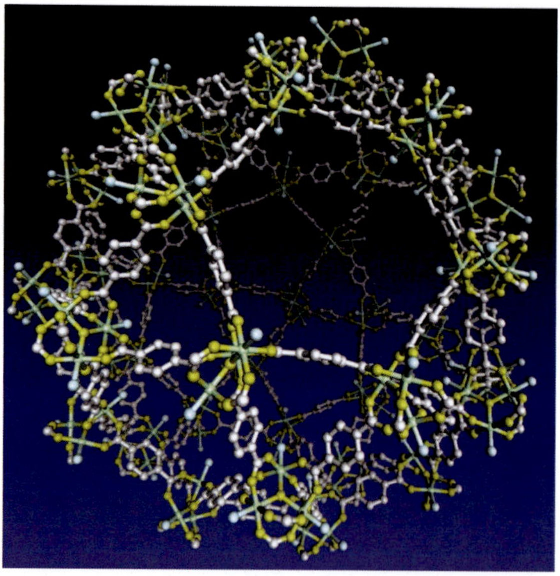

Plate 4 Structure of a hybrid material obtained by predictive hydrothermal synthesis (diameter 34 Å, volume 20 600 $Å^3$, hexagonal window 16.5 Å). Reprinted with permission from *Chemical Society Reviews*, Hybrid porous solids: past, present, future by Gerard Ferey, 37, 1, 191–214. Copyright (2008) RSC

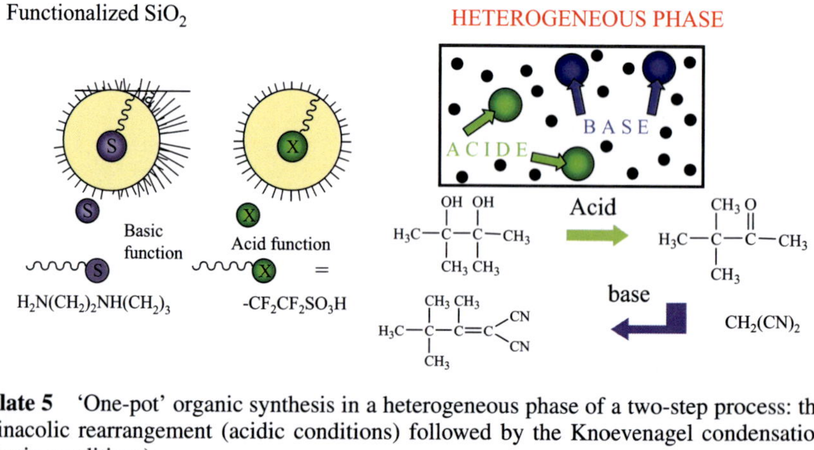

Plate 5 'One-pot' organic synthesis in a heterogeneous phase of a two-step process: the pinacolic rearrangement (acidic conditions) followed by the Knoevenagel condensation (basic conditions)

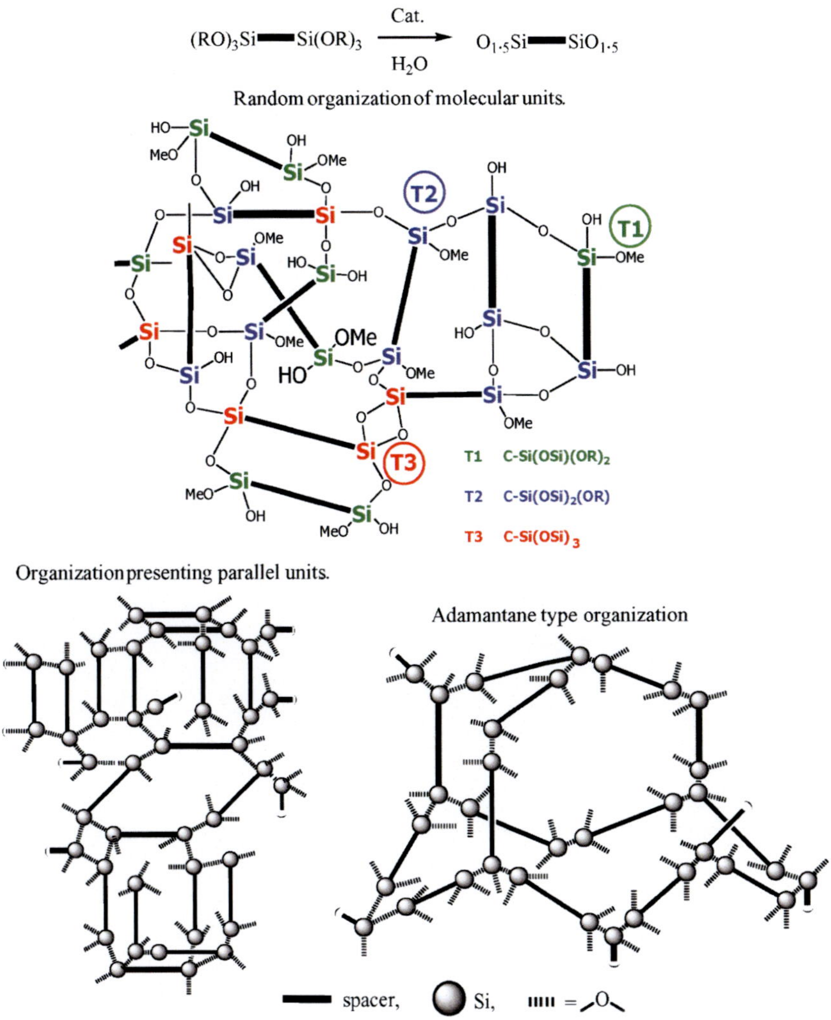

Plate 6 Three different types of arrangement are given as possible examples of organization of the organic units in the material. The organizations are different but all the data (weight analysis, BET, porosity, IR, NMR, etc.) are the same. Reprinted with permission from *New J. Chem.*, Supramolecular self-organization in non-crystalline hybrid organic-inorganic nanomaterials induced by van der Waals interactions, Lerouge, Frédéric; Cerveau, Geneviève; Corriu, Robert J.P., *30*, 1364–1376. Copyright (2006) RSC

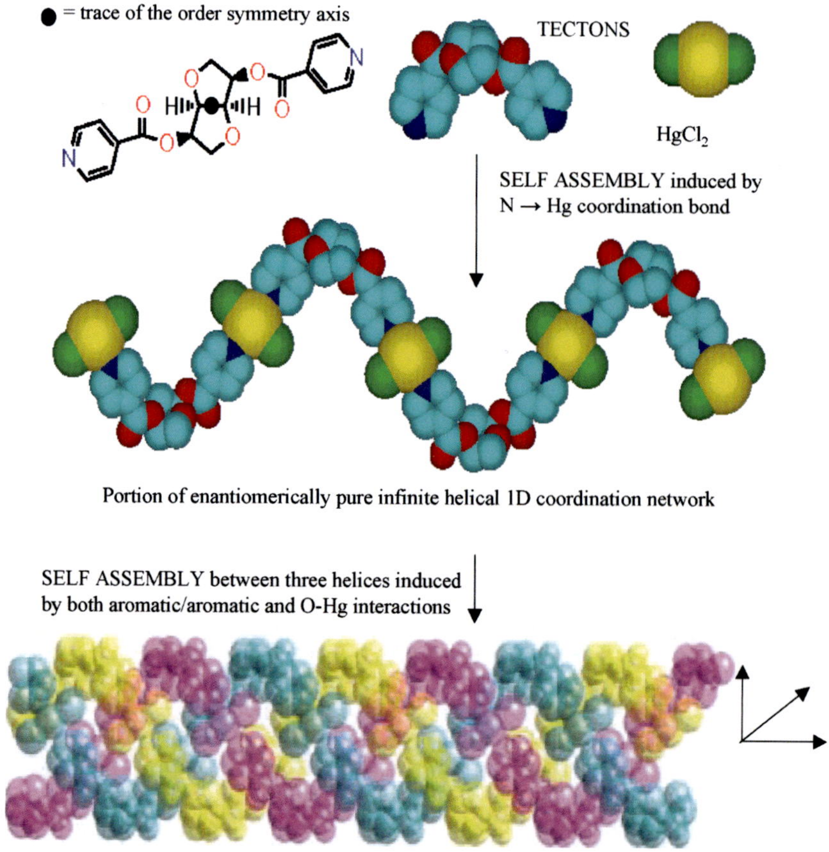

Plate 7 Formation of an enantiomerically pure, triple-stranded helical coordination network. Reprinted with permission from *New J. Chem.*, Supramolecular self-organization in non-crystalline hybrid organic-inorganic nanomaterials induced by van der Waals interactions, Lerouge, Frédéric; Cerveau, Geneviève; Corriu, Robert J.P., *30*, 1364–1376. Copyright (2006) RSC

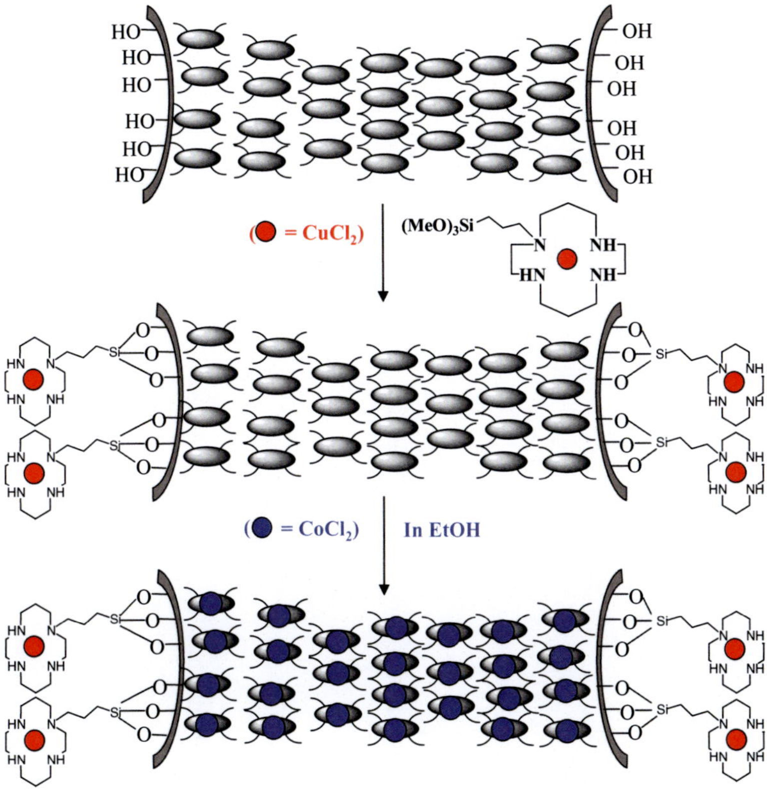

Plate 8 Introduction of transition metal ions (Co^{2+}, Cu^{2+}) either in the walls or in the pores of the mesoporous material shown in Figure 6.11. Reprinted with permission from *New Journal of Chemistry*, Control of coordination chemistry in both the framework and the pore channels of mesoporous hybrid materials, Corriu, Robert J. P.; Mehdi, Ahmad; Reye, Catherine; Thieuleux, Chloe, *27*, 905–908. Copyright (2003) RSC

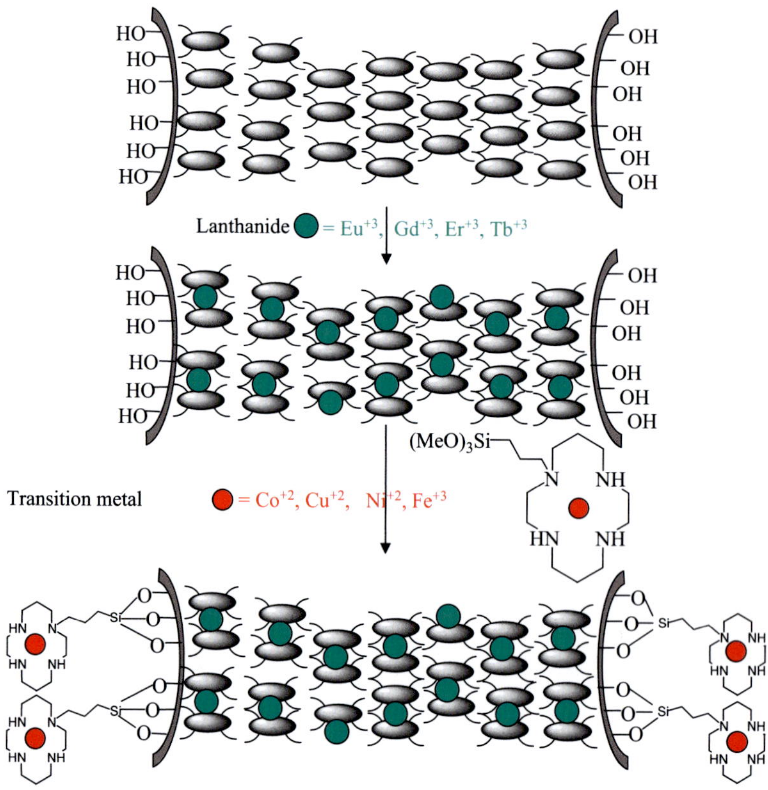

Plate 9 Introduction of lanthanides in the framework (one ion for two chelating units) and fixation of transition metals in the pores

Table 5.2 Comparison between the calculated and experimental values for the Si...Si distance from precursors **1–5** (**cf. Figure 5.26**)

		Values (Å)				Values (Å)	
		Calc.	Exp.			Calc.	Exp.
1	n = 1	7.8	7.8	4	m = 1	11.5	11.4
2	n = 2	12.3	12.5	5	m = 2	15.5	15.7
3	n = 3	15.7	16.3				

Reprinted with permission from *New J. Chem.*, Supramolecular self-organization in non-crystalline hybrid organic-inorganic nanomaterials induced by van der Waals interactions, Lerouge, Frédéric; Cerveau, Geneviève; Corriu, Robert J.P., *30*, 1364–1376. Copyright (2006) RSC

tal conditions (solvent, catalyst, concentration of the reagents, etc.) but their position remains unchanged. Again, we come across kinetic control, previously encountered when considering texture.

This densification of matter occurs with the progressive disappearance of H_2O molecules, and the concomitant appearance of alcohol molecules. The organic entities are in favorable conditions for creating a self-organization. *Indeed, formation of Si–O–Si bonds transforms the intermolecular interactions into intramolecular interactions, thus favoring a decrease in entropy.* This decrease in entropy permits the organization between the organic spacers thanks to weak lipophilic van der Waals type interactions whose energy is too low to induce an organization in solution (Figure 5.27).

Organization at the micrometric scale has been studied by birefringence, which measures the influence of the material on a polarized white light.[51–53] The detection of birefringence points towards an *anisotropic organization of the material at a micrometric scale*, which corresponds to the wavelength range of the white light used for measurements.

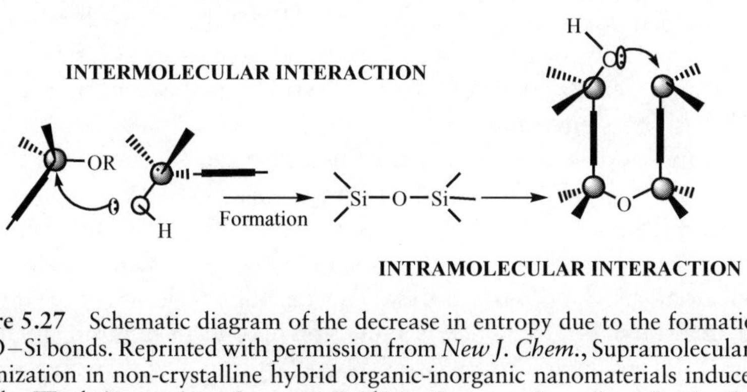

Figure 5.27 Schematic diagram of the decrease in entropy due to the formation of Si–O–Si bonds. Reprinted with permission from *New J. Chem.*, Supramolecular self-organization in non-crystalline hybrid organic-inorganic nanomaterials induced by van der Waals interactions, Lerouge, Frédéric; Cerveau, Geneviève; Corriu, Robert J.P., *30*, 1364–1376. Copyright (2006) RSC

Figure 5.28 Birefringence photographs observed after hydrolysis of **4** which reacts without catalyst. Reprinted with permission from *New J. Chem.*, Supramolecular self-organization in non-crystalline hybrid organic-inorganic nanomaterials induced by van der Waals interactions, Lerouge, Frédéric; Cerveau, Geneviève; Corriu, Robert J.P., *30*, 1364–1376. Copyright (2006) RSC

Figure 5.28 shows some birefringence photographs taken with appropriate observation cells. The formation of cracks is observed in almost every case. The cracks are due to the shrinkage following the polycondensation at silicon. The formation of the Si–O–Si bonds changes the precursor molecules (monomers), which are isolated in solution, into a denser medium, finally giving a solid in which all the silicon atoms are linked by siloxane bridges. Let us recall that the polycondensation yield is 100% (*cf.* Section 3.5).

Thus, black isotropic regions corresponding to voids are observed between birefringent chunks of gel. The voids are due to shrinkage following the polycondensation reaction, which transforms a solution into a much more condensed solid phase. All the chunks are birefringent. The black parts inside the chunks correspond to the traces of the polarizer and the analyzer. Rotation of the sample still gives the same image of the chunk and of its environment.

Dynamic studies have been performed in order to determine during which step of the sol-gel process birefringence occurs.[51] Experiments were realized with a controlled partial pressure of water, using open cells containing the pure crystalline precursor (**4** and **5**, Figure 5.26) which can be polycondensed *without catalyst*. This peculiarity permits the study of the formation of birefringence as a function of time, by exposing a sample of the solid precursor to an atmosphere saturated with water vapor. Figure 5.29 shows the slow appearance of birefringence after the sol–gel

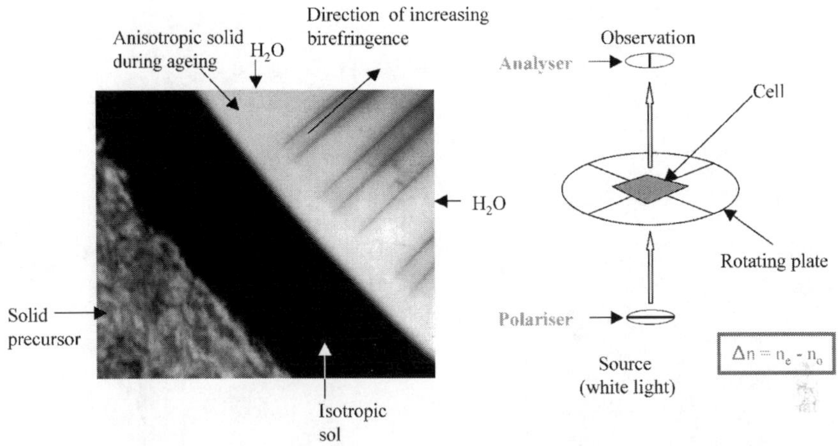

Figure 5.29 Dynamic representation of birefringence formation and a schematic diagram of birefringence measurement. Reprinted with permission from *New J. Chem.*, Supramolecular self-organization in non-crystalline hybrid organic-inorganic nanomaterials induced by van der Waals interactions, Lerouge, Frédéric; Cerveau, Geneviève; Corriu, Robert J.P., *30*, 1364–1376. Copyright (2006) RSC

transition. *It can therefore be concluded that the extension of the organization from the nanometric to the micrometric scale occurs during ageing of the solid.*

The isotropic dark region corresponds to the formation of the sol after hydrolysis of the solid precursor. The hydrolytic polycondensation continues as the water vapor progresses into the cell: the inner regions, less accessible, start to hydrolyze while the edges of the cell are already polycondensed. Figure 5.29 clearly shows that the sol is amorphous while the birefringence appears during the formation of the solid and persists throughout its transformation.[51] This transformation is shown schematically in Figure 5.30.

In conclusion, this study shows that the hydrolytic polycondensation leads to two types of organization: one at the nanometric scale, revealed by X-ray diffraction, which occurs during the polymerization in solution; and another at the micrometric scale, shown by birefringence studies, occurring in the solid phase during the ageing process.

5.6.4 Generalization of the Self-Organization Phenomenon

It has clearly been demonstrated that the self-organization phenomenon is general and is not restricted to the case of linear spacers. Figure 5.31 shows

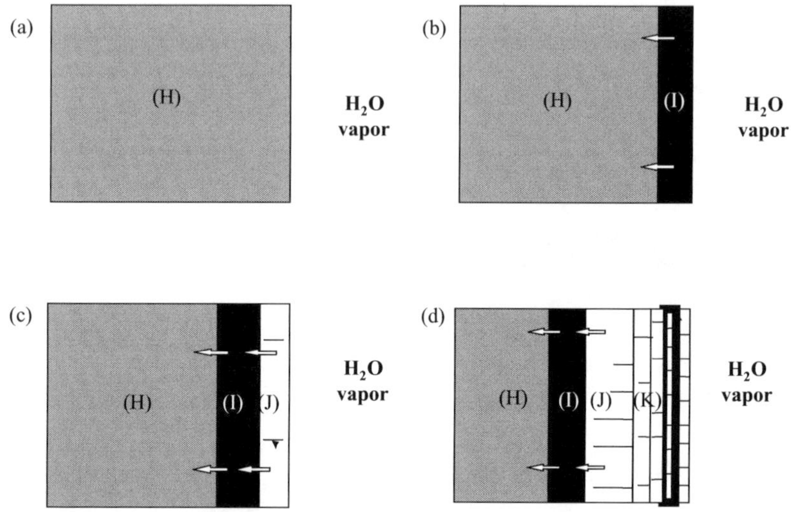

Figure 5.30 Birefringence experiments: Modifications of part of a cell during the polycondensation process: (a) precursor; (b) sol formation; (c) birefringence; and (d) changes in birefringence during ageing. Reprinted with permission from *New J. Chem.*, Supramolecular self-organization in non-crystalline hybrid organic-inorganic nanomaterials induced by van der Waals interactions, Lerouge, Frédéric; Cerveau, Geneviève; Corriu, Robert J.P., *30*, 1364–1376. Copyright (2006) RSC

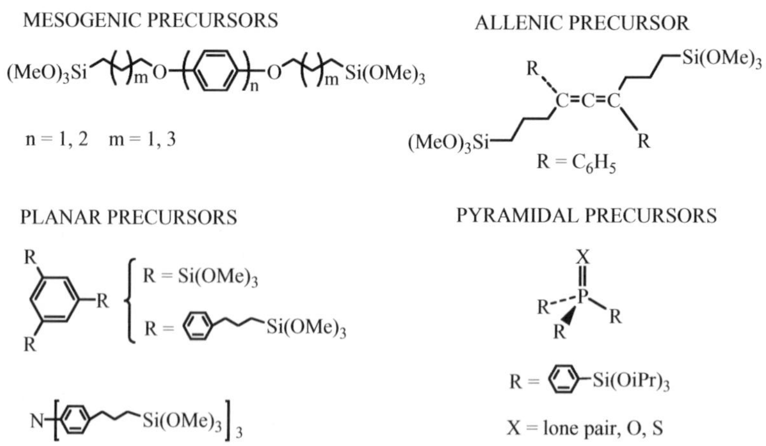

Figure 5.31 Molecular precursors leading to materials organized at the nanometric (X-ray) and micrometric (birefringence) scales. Reprinted with permission from *New J. Chem.*, Supramolecular self-organization in non-crystalline hybrid organic-inorganic nanomaterials induced by van der Waals interactions, Lerouge, Frédéric; Cerveau, Geneviève; Corriu, Robert J.P., *30*, 1364–1376. Copyright (2006) RSC

some of the systems presenting different geometries that have been studied.[56–59]

All the systems studied up to now have similar characteristics: (1) an X-ray diffraction pattern exhibiting signals characteristic of the molecular spacer and indicative of a nanometric scale organization; (2) emergence of birefringence during ageing of the solid; (3) formation of cracks corresponding to the transition from a colloidal solution of crosslinked polymers to a polycondensed solid; (4) observation of a wide variety of birefringence images, which depend strongly on the nature of the precursors; (5) kinetic control of the diffraction patterns; (6) orientation of the optical axis strongly dependent on the geometry of the precursor. This axis is generally perpendicular to the cracks, with one exception (allenic precursors).[56] Points (2)–(6) correspond to different organizations at the micrometric scale.

Let us recall how the optical axis is defined and what its characteristics are. *The optical axis is a purely experimental datum*, measured with the Berek compensator. In liquid crystals, it shows the orientation of molecules at the micrometric scale.[60,61] In polycondensed systems, it can be assumed that this experimental datum corresponds to the orientation of the micrometric-sized colloids when they aggregate during gelation. Figure 5.32 shows an outline of the organization process.

The case of the aliphatic precursors $(RO)_3Si(CH_2)_nSi(OR)_3$ is particularly interesting. After polycondensation, although these materials exhibit X-ray diffraction patterns, no birefringence is observed.[52] Such a result can be explained by the hydrophobic behavior of aliphatic systems placed under the hydrophilic conditions required for hydrolytic polycondensation. During the process, the hydrophobic $Si(OR)_3$ groups are transformed

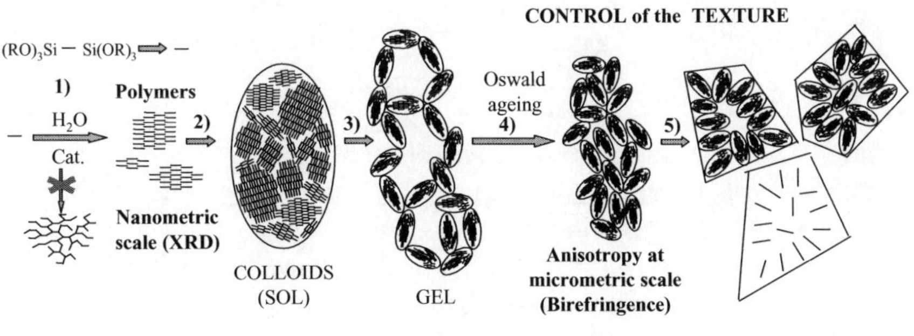

Figure 5.32 Control of the self-organization of nanostructured hybrid materials. A schematic diagram of the different steps involved in the process (visualization of the optical axis on a chunk). Reprinted with permission from *New J. Chem.*, Supramolecular self-organization in non-crystalline hybrid organic-inorganic nanomaterials induced by van der Waals interactions, Lerouge, Frédéric; Cerveau, Geneviève; Corriu, Robert J.P., *30*, 1364–1376. Copyright (2006) RSC

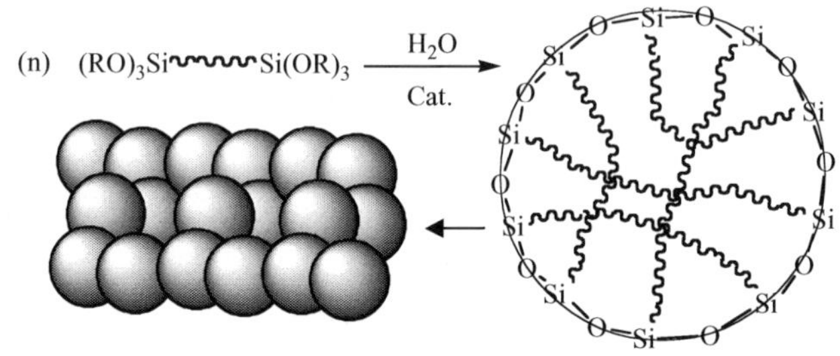

Figure 5.33 Globular-type aggregates obtained from aliphatic systems. Reprinted with permission from *New J. Chem.*, Supramolecular self-organization in non-crystalline hybrid organic-inorganic nanomaterials induced by van der Waals interactions, Lerouge,Frédéric; Cerveau, Geneviève; Corriu, Robert J.P., *30*, 1364–1376. Copyright (2006) RSC

into hydrophilic $Si(OR)_2OH$ and $SiOR(OH)_2$ groups. Thus, the flexible hydrophobic aliphatic chains amalgamate into a spheroidal micelle-type geometry. Finally, the stacking of these more or less distorted spheres occurs during ageing. A solid build up by such a stacking presents a very weak anisotropy, which explains the absence of birefringence (Figure 5.33). However, the micrometric organization is detectable by X-ray diffraction.

5.6.5 Study of Tetrahedral Systems

Tetrahedral structures,[62] which favor isotropy (there is no liquid crystal presenting this type of geometry), have been used to check the generality of this type of organization. Molecules having a tetrahedral core and bearing eight long aliphatic groups $[O(CH_2)_9CH_3]$, explored as possible mesophases[§] do not present any birefringence properties, proving that the lipophilic interactions between the eight $[O(CH_2)_9CH_3]$ chains are not enough to induce an *intermolecular self-organization* of the tetrahedral patterns.[63] It is interesting to note that two shorter aliphatic chains can induce birefringence on linear systems (Figure 5.34). The aromatic core does not present any birefringence properties, however, the presence of two aliphatic chains having five CH_2 units is sufficient to induce smectic or nematic liquid crystal type anisotropic organization.[64] This comparison clearly demonstrates that *a tetrahedral geometry does not favor anisotropy.*

[§]A mesophase can induce an organization detectable by birefringence.

(a)

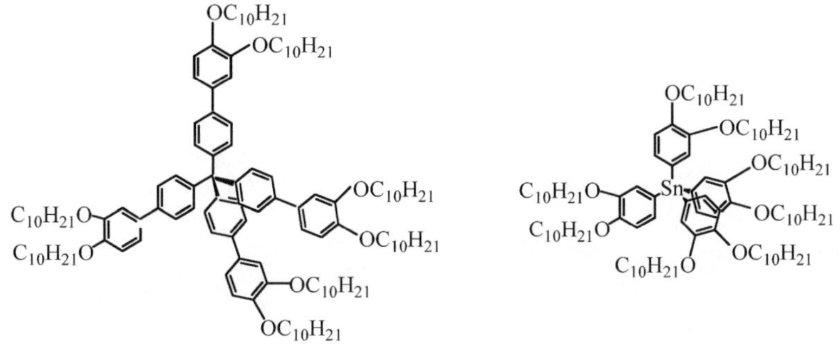

(b)

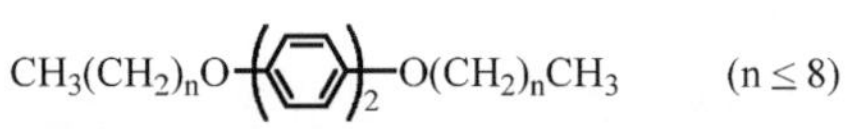

Figure 5.34 (a) Tetrahedral molecules without liquid crystal properties. (b) Linear molecular precursors presenting liquid crystal behavior. Reprinted with permission from *New J. Chem.*, Supramolecular self-organization in non-crystalline hybrid organic-inorganic nanomaterials induced by van der Waals interactions, Lerouge, Frédéric; Cerveau, Geneviève; Corriu, Robert J.P., *30*, 1364–1376. Copyright (2006) RSC

Studies have been carried out on tetrahedral precursors having the same $(C_6H_4)_4Sn$ or $(C_6H_4)_4Ge$ basic structure, substituted by $Si(OR)_3$ groups in the para position of the metal center.[62] Despite the fact that they contain no mesogenic group, these tetrahedral precursors, when polycondensed, give rise to an organization at the nanometric scale and another at the micrometric scale. The existence of the nanometric organization is demonstrated by X-ray diffraction patterns showing clear diffraction patterns. The broad diffraction signals observed correspond, without any doubt, to the existence of a nanometric order in the materials. For instance, in the case of the tin compound, the strongest signal at ~7 Å can be correlated with values obtained by molecular modeling, and corresponds to the distance separating the mean planes of tin atoms on the one hand and siloxane bridges on the other. The $SiO_{1,5}$ network corresponds to the signal at ~4 Å.

Birefringence measurements show that the size and arrangement of the rod-like chunks is regular. The optical axes are oriented perpendicular to the edges of the chunks (Figure 5.35).

This corresponds to an orientation of the organized matter at the micrometric scale. In the case of liquid crystals it is connected with the orientation of independent molecules anisotropically organized. These results observed in the case of tetrahedral units clearly show that the self-organization process is a general one since it occurs even when the organic precursor presents a geometry that favors isotropy. *Furthermore the regularity of size and distribution extends from the micrometric to centimetric scale.*

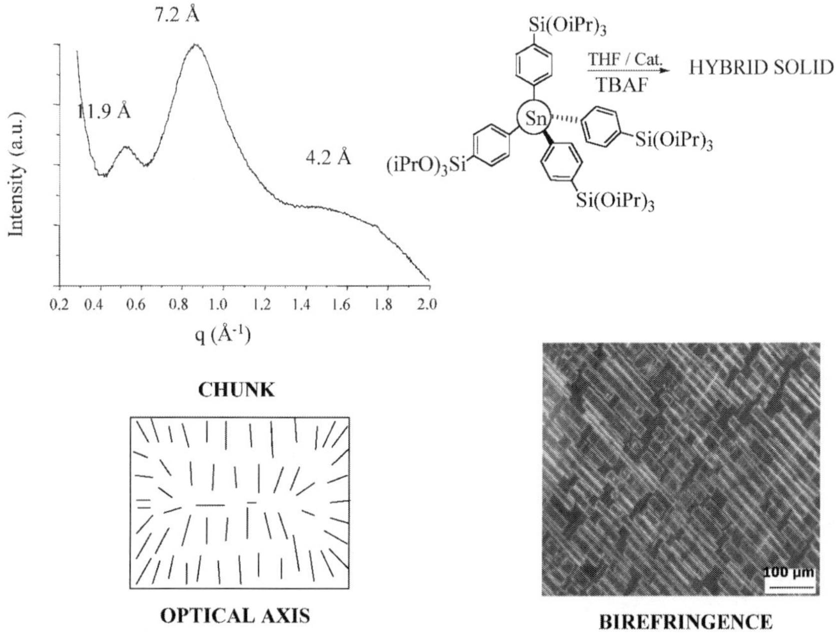

Figure 5.35 X-ray diagram, birefringence photograph and schematic diagram of the optical axis obtained with a tetrahedral precursor. Reprinted with permission from *New J. Chem.*, Supramolecular self-organization in non-crystalline hybrid organic-inorganic nanomaterials induced by van der Waals interactions, Lerouge, Frédéric; Cerveau, Geneviève; Corriu, Robert J.P., *30*, 1364–1376. Copyright (2006) RSC

This example demonstrates clearly that a high level of organization can be reached by hydrolytic polycondensation. The formation of Si–O–Si bonds drastically minimizes entropy and thus allows the self-organization of nonmesogenic units at three levels: nanometric, micrometric and millimetric.

Organization at the micrometric scale has been observed in all cases: linear and planar,[57,65–68] tetrahedral,[62] and even orthogonal geometries.[56] In conclusion, organization of the molecular entities during the polycondensation process can be considered a general phenomenon.

5.6.6 Kinetic Control of Self-Organization

The self-organization phenomenon is a kinetically controlled process which is greatly influenced by experimental factors (solvent, catalyst, concentration of the reagents, temperature, etc.). In fact, every experimental

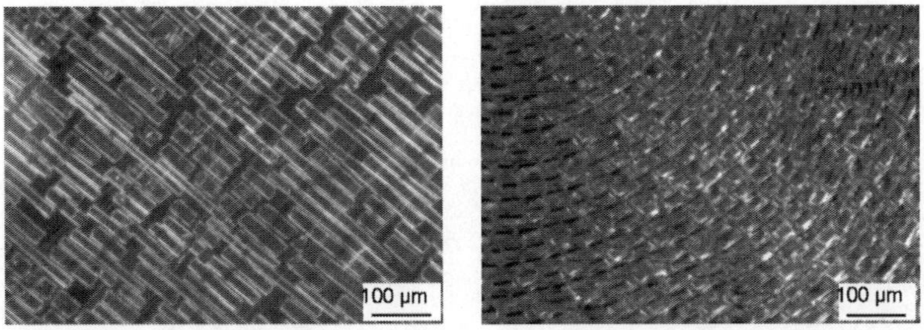

$\Delta n = 1.7 \times 10^{-3}$
TBAF

$\Delta n = 2.1 \times 10^{-3}$
HCl

Figure 5.36 Influence of the catalyst on the micrometric and millimetric organizations in the case of a tin tetrahedral precursor. Reprinted with permission from *New J. Chem.*, Supramolecular self-organization in non-crystalline hybrid organic-inorganic nanomaterials induced by van der Waals interactions, Lerouge, Frédéric; Cerveau, Geneviève; Corriu, Robert J.P., *30*, 1364–1376. Copyright (2006) RSC

parameter affects the intensity of the X-ray diffraction signals, the position of the signals remaining unchanged. It also influences the birefringence phenomenon (form and size of chunks, intensity, orientation of the optical axis). This can be illustrated by some examples (Figure 5.36). Clearly, the birefringence obtained in the case of tetrahedral geometries with HCl or TBAF as catalysts is not identical.[62] However, the main characteristics are the same: the birefringence intensities are very similar, the millimetric organization and the directions of the optical axes are analogous. In both cases, the X-ray patterns are comparable.

More important modifications have been observed. Thus, in the case of an allenic precursor, presenting a twisted geometry, a noteworthy effect of the temperature has been observed.[56] The X-ray diffraction patterns and the birefringence images vary drastically, as seen in Figure 5.37. The X-ray pattern exhibits, as usual, two broad signals at room temperature (20 °C). At low temperature (3 °C), additional weak signals, indicative of a better nanometric scale organization are observed. At 20 °C, the birefringence figures show the presence of parallel wavy cracks that propagate throughout the whole cell (2 cm × 2 cm) over distances of several millimeters. Such a phenomenon has never been observed before. The optical axis is oriented parallel to the edges of the cracks (Figure 5.37a). In contrast at 3 °C, the birefringence appears without cracks and the optical axis is oriented radially around a central point (Figure 5.37b).

Experience thus shows that self-organization and texture are both under kinetic control. This is *not* a handicap, since it is possible to optimize the physical characteristics of the material by using the kinetic

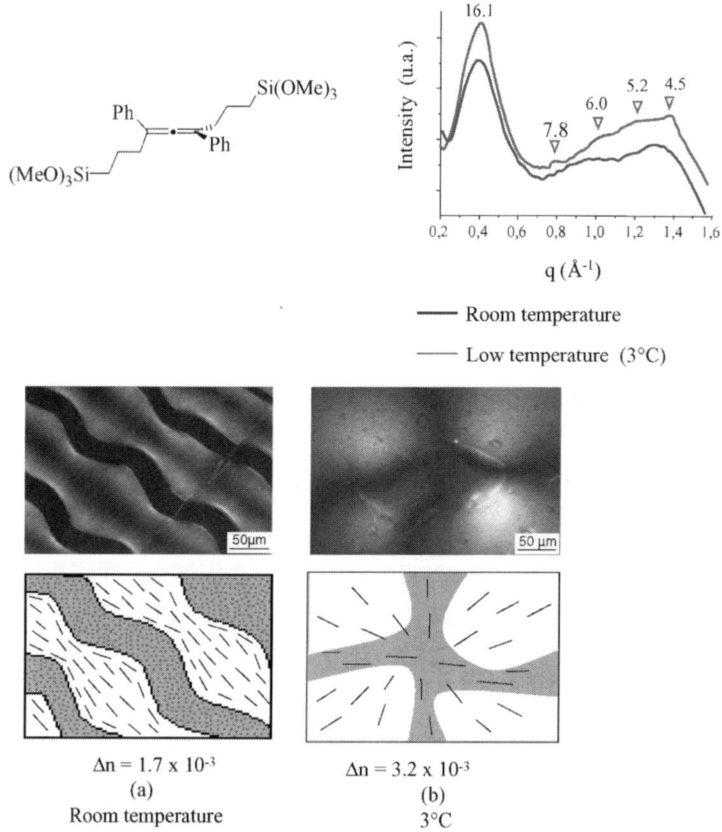

Figure 5.37 Influence of temperature on nanometric scale organization (X-ray) and on micrometric scale organization (birefringence). Reprinted with permission from *New J. Chem.*, Supramolecular self-organization in non-crystalline hybrid organic-inorganic nanomaterials induced by van der Waals interactions, Lerouge, Frédéric; Cerveau, Geneviève; Corriu, Robert J.P., *30*, 1364–1376. Copyright (2006) RSC

parameters: solvent, hydrophilic/lipophilic balance, temperature, nature and concentration of the reagents, of the substrates and of the catalyst. Thus, *the texture and organization can be adjusted during the inorganic polymerization; the stabilization of the material being performed by a hydrothermal treatment in relatively mild conditions* (<150 °C, low concentration of acid or NH_4F nucleophilic catalyst). This treatment stops further changes in the solid by preventing polycondensation at silicon, as well as the redistribution of Si–O–Si bonds. All this happens without significant modification of the material's characteristics. The solid then becomes stable in the normal conditions of utilization.

Texture and structure are thus both sensitive to the same parameters, but in different ways, since the characteristics involved are not the same. Texture control corresponds to the organization of the solid external surface (specific surface area and porosity), while structural control concerns the relative arrangement of the organic units within the material.

In the process that goes from a homogeneous solution to a solid, the turning point appears to be the sol–gel transition (Figure 5.32, step 3) and the different steps occcuring before and after this point must be considered.

The first phase of the organization occurs in solution, during the mainly irreversible Si–O–Si bond formation (Figure 5.32, steps 1 and 2). Oligomers form and grow giving polymers by polycondensation. Organic units are brought close to each other and supramolecular van der Waals type interactions are favored, leading to nanometric scale self-organization. During these steps, the nanometric organization between the organic spacers is built up by formation of aggregates of different sizes. For this reason, the X-ray diffraction patterns exhibit broad signals instead of Bragg peaks. These aggregates, which are formed of the same elemental 'bricks', link covalently to form micrometric-sized colloids, constituting the sol, a viscous liquid. Thus in solution, during the first steps of the polycondensation, the van der Waals type interactions between the organic units induce a short range order over nanometric domains.

The formation of covalent bonds between colloids corresponds to the sol–gel transition (Figure 5.32, step 3).

The second important phase of the process concerns the changes that occur in the solid state, during ageing (Figure 5.32, steps 4 and 5). The texture and organization of the solid strongly depend on the parameters that control the ageing of the material in the solid state. *The birefringence phenomenon and the orientation of the optical axis are the result of the transformations that take place in the solid state.* The densification of the Si–O–Si framework, the syneresis step (expulsion of the solvent), and the reorganization of the micrometric aggregates generate stresses that are released by formation of cracks or regions corresponding to shrinkage of the material. As demonstrated by Raman studies, the observed cracks correspond to voids. They lead to the birefringence phenomenon due to the propagation of the short range order to domains larger than the wavelength of light. The orientation of the optical axis corresponds to an orientation of the organized matter at the micrometric scale.

5.6.7 Some Reflections on the Observed Self-Organization

Nanostructured solids obtained from hydrolytic polycondensation are not badly organized crystalline systems or crystalline organizations over short distances. *They correspond to a structural mode completely different from that of crystals*. In a crystalline arrangement, the crystal lattice reproduces itself identically in each of the three directions of space. It can be different from one direction to another in lamellar systems or in unidirectional-type crystals. However, *the organization of the elemental lattice is the same in each direction of space*. In addition, crystal growth is under thermodynamic control.

Thus, a cubic crystal of NaCl has a lattice replicating itself identically in the three directions of space. It is the same with crystals of any lattices: quadratic, monoclinic, etc. In lamellar crystals (e.g. mica or clay), the crystal has two identical directions corresponding to the plane of lamellae. The third direction, perpendicular to the two others, corresponds to the stacking of identical lamellae (Figure 5.38).

The case of unidirectional crystals is less usual than the two previous examples. It will be illustrate here by the case of a three-fold helix obtained by organometallic synthesis (Figure 5.39).[69]

Using an enantiomerically pure organic tecton derived from isomannide, the formation of an enantiomerically pure helical coordination network was achieved and structurally characterized by X-ray diffraction

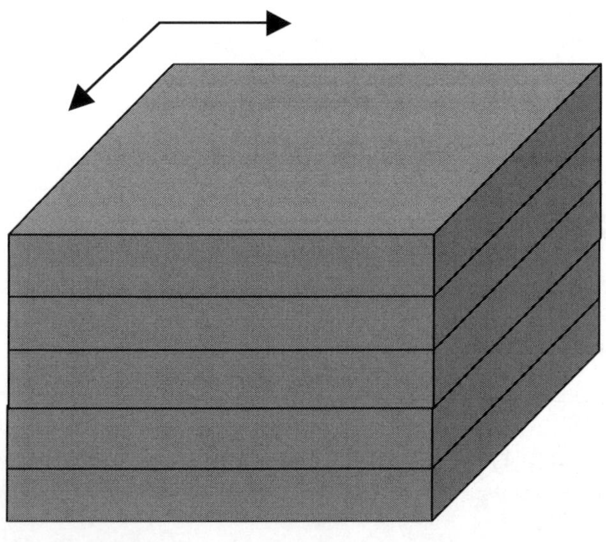

Figure 5.38 Lamellar-type crystal

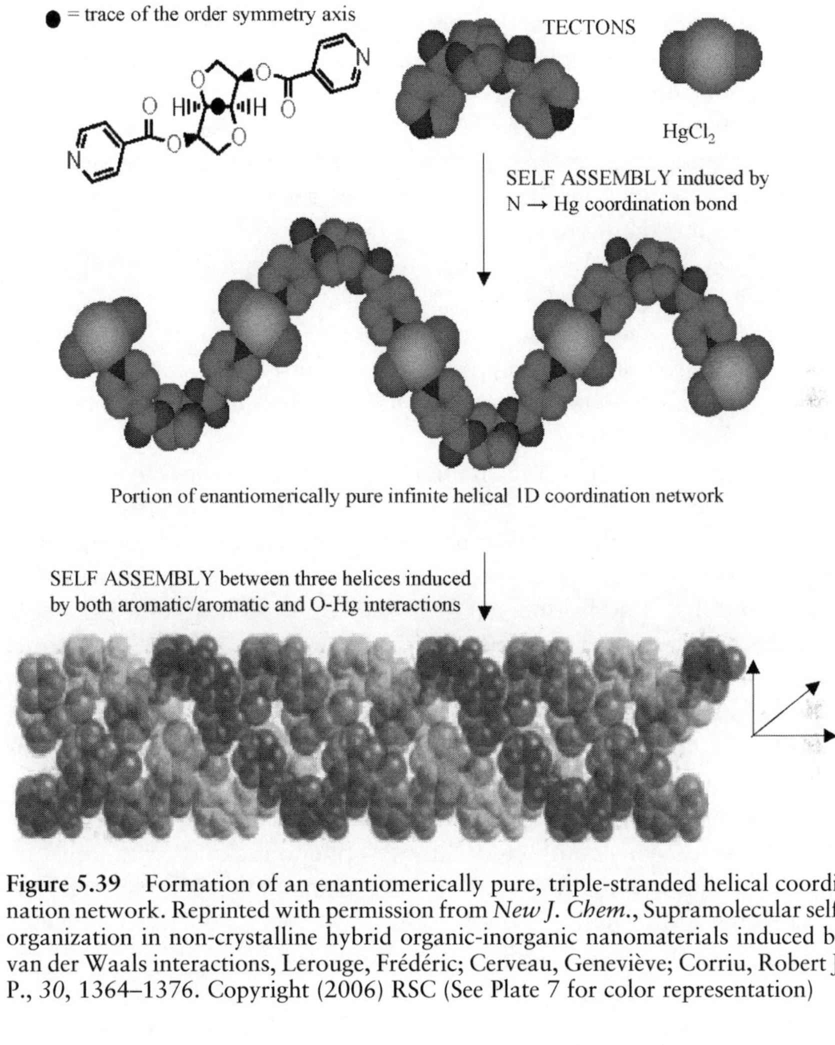

Figure 5.39 Formation of an enantiomerically pure, triple-stranded helical coordination network. Reprinted with permission from *New J. Chem.*, Supramolecular self-organization in non-crystalline hybrid organic-inorganic nanomaterials induced by van der Waals interactions, Lerouge, Frédéric; Cerveau, Geneviève; Corriu, Robert J. P., *30*, 1364–1376. Copyright (2006) RSC (See Plate 7 for color representation)

on a single crystal. Three infinite helical strands self-assemble into a triple-stranded noncylindrical architecture through aromatic/aromatic interactions. The triple-stranded helices are packed parallel to each other and further interconnected through O–Hg interactions. The overall structure may be described as a three-dimensional network resulting from the interconnection of triple-stranded helical one-dimensional networks. Such organization results from favorable van der Waals interactions.

In this example of a linear crystal, one of the axes lies in the direction of the crystal. The two others correspond to the stacking of the triple unidirectional helices.

The organization observed in nanostructured hybrid materials, which strictly speaking are amorphous systems, results from a process completely different from crystalline growth. Indeed, in the case of crystals, an equilibrium develops between the solid and the solution. *Crystal growth occurs by integration of the constitutive building blocks and according to the crystal organization.* In contrast, during the hydrolytic polycondensation described here, the phases of organization are different, depending on the progress of the process. This type of organization has already been described (*cf.* Section 5.6.6).

To conclude, we can say that the organization occurring during the hydrolytic polycondensation is *an arrangement mode completely different from crystal growth, which, however, can organize matter on a large scale.*

The formation of Si–O–Si bonds forces the molecules closer to each other and imposes the geometry. This minimizes the entropy, favoring intramolecular van der Waals type interactions that can only occur when organic groups are sufficiently close, since their energy is inversely proportional to the distance according to the R^{-6} law. Thus, the self-organization described is not spontaneous. *It arises from the close contact that is imposed on precursor molecules during polycondensation in the liquid phase.* It is responsible for the organization at the nanometric scale.

After the sol–gel transition, the system is governed by interactions between micrometric-sized particles that become connected. The formation of both texture and structure is controlled by the modifications in the solid state. This completely different phenomenon is controlled by the surface functionalities and by the propagation of cracks. It is responsible for the solid characteristics as the polycondensation occurs and the Si–O–Si bonds redistribute during the Oswald ageing. In the present state of research, it is difficult to analyze further. The texture indicates the state of the solid surface while the birefringence measures the organization of the colloidal entities which constitute the solid after polycondensation.

5.7 LAMELLAR MATERIALS

In the previous examples, the experimental conditions are essentially hydrophobic; the H_2O concentration is limited to the stoichiometric amount required for the total polycondensation. It was interesting to find out what kinds of materials could be obtained in hydrophilic conditions. The polycondensation has been therefore studied in aqueous medium at pH = 1.5. This pH corresponds to the isoelectric point which, in the case of $Si(OR)_4$, corresponds to a suspension of polysilicic acid $[Si(OH)_4]_n$. In these conditions, the materials obtained are spectacularly different from those obtained in the previous experiments (stoichiometry in H_2O) with the same precursors.

For example, the hydrophilic polycondensation of α,ω-bis(trimethoxysilyl)alkanes leads to lamellar materials in the case of C_{12} and C_{18} chains.[70] Figure 5.40 shows the X-ray diffraction pattern of the material obtained with a C_{12} precursor. It is characteristic of a lamellar organization. Thus, a sharp and intense peak corresponding to the thickness *d* of the lamellae, as well as the harmonic peaks at *d*/2, *d*/3 and *d*/4 are observed. This thickness *d* corresponds to the length of the precursor (20.4 Å).

Such an organization is not observed if the hydrolytic polycondensation of the same precursor is carried out in the presence of a stoichiometric amount of H_2O[70] (Figure 5.33). This difference in behavior can be explained by the hydrophobicity of the aliphatic chains, which tend to aggregate together in a hydrophilic medium. Indeed, hydrolysis transforms the $-Si(OR)_3$ groups into $Si(OR)_{3-x}OH_x$ polar heads. The polycondensation at silicon leads to the formation of lamellar systems by forcing the regrouping of the hydrophilic heads on the one hand and that of the hydrophobic chains on the other hand.

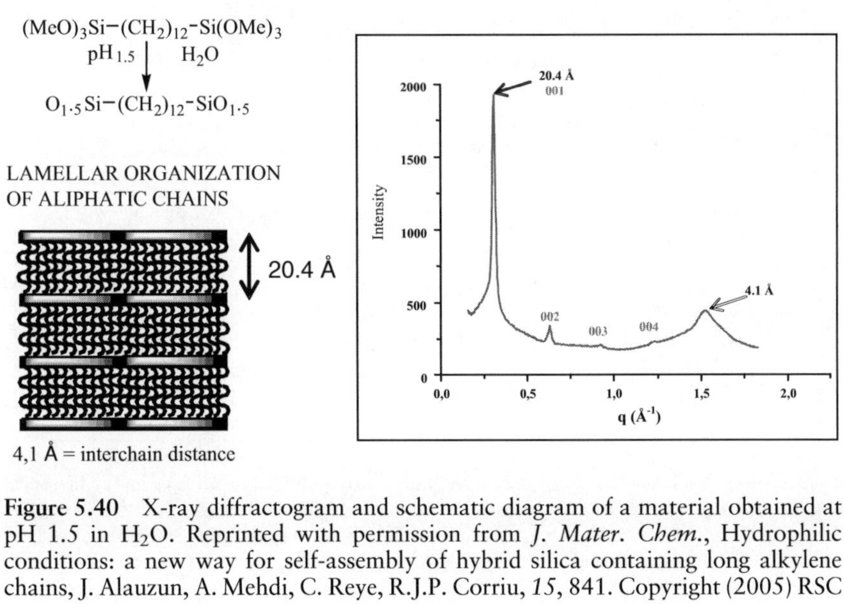

Figure 5.40 X-ray diffractogram and schematic diagram of a material obtained at pH 1.5 in H_2O. Reprinted with permission from *J. Mater. Chem.*, Hydrophilic conditions: a new way for self-assembly of hybrid silica containing long alkylene chains, J. Alauzun, A. Mehdi, C. Reye, R.J.P. Corriu, *15*, 841. Copyright (2005) RSC

In order to test the generality of this explanation, a very hydrophobic system $(MeO)_3Si-(CH_2)_{30}Si(OMe)_3$ has been examined. Hydrolysis carried out in the same conditions as for the C_{12} and C_{18} precursors leads to a hexagonal phase. The hydrophobicity of the C_{30} chain forces the precursor to behave as a surfactant in a hydrophilic medium. The aliphatic chains then gather together leading to a micellar medium. The high viscosity of the sol induces a hexagonal phase by regrouping of the individual micelles (Figure 5.41).

The X-ray diffraction pattern of the material corresponds to a hexagonal structure of parallel cylinders linked together by Si–O–Si bonds. The first signal at $d = 35.7$ Å corresponds to the distance between the axes of two neighboring cylinders. The hexagonal phase is proved by the presence of two satellite peaks located at $d/\sqrt{3}$ and $d/2$.

This methodology offers interesting prospects. It is in fact possible to introduce potential functions, for example disulfide bridges, on a long aliphatic chain.

Disulfide bridges are lipophilic. Using α,ω-silylated precursors of $(MeO)_3(CH_2)_nS-S(CH_2)_nSi(OMe)_3$ type and a method identical to that

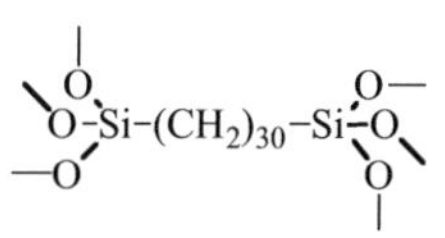

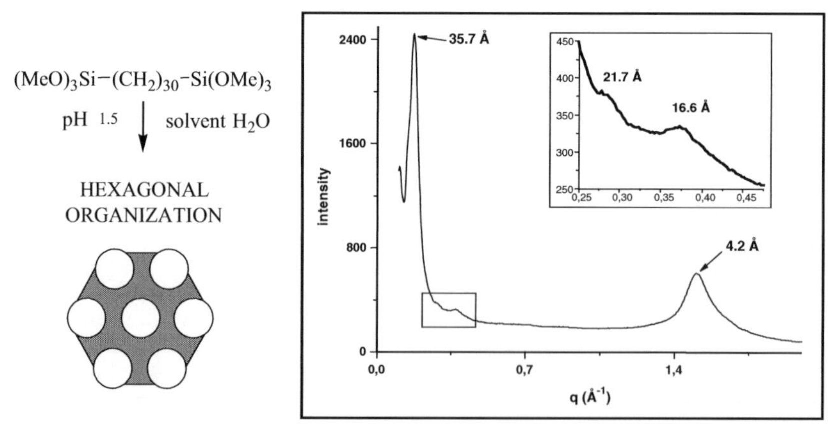

Figure 5.41 Schematic diagram of hydrolysis of $(MeO)_3Si\text{-}(CH_2)_{30}Si(OMe)_3$ precursor and the X-ray diffractogram of the hexagonal phase. Reprinted with permission from *J. Mater. Chem.*, Hydrophilic conditions: a new way for self-assembly of hybrid silica containing long alkylene chains, J. Alauzun, A. Mehdi, C. Reye, R.J.P. Corriu, *15*, 841. Copyright (2005) RSC

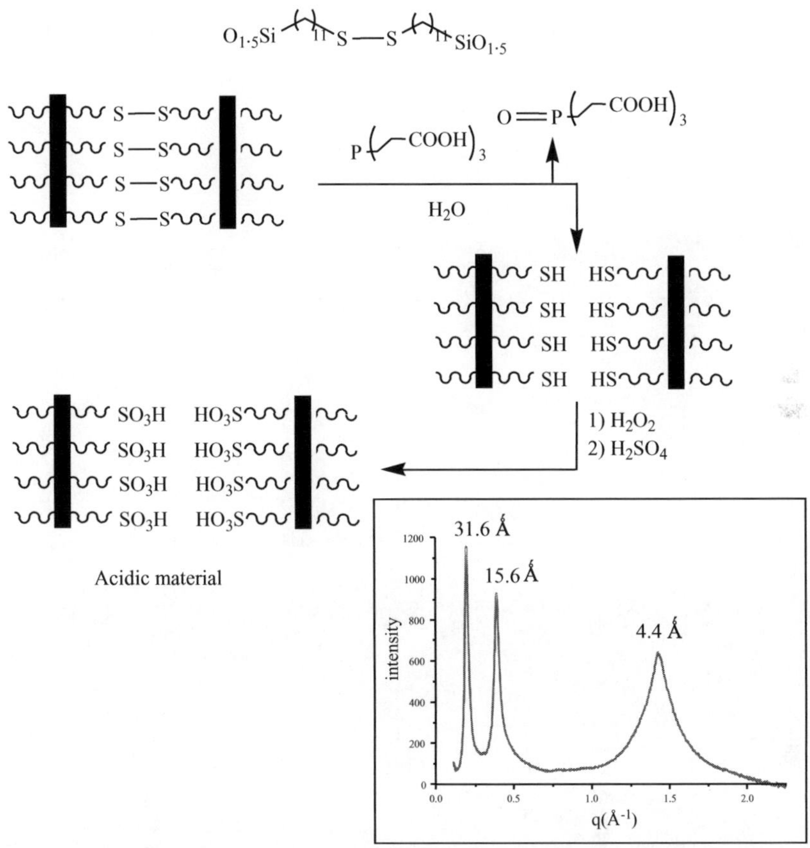

Figure 5.42 Synthesis of functional lamellar materials from $(MeO)_3Si(CH_2)_{11}S–S(CH_2)_{11}Si(OMe)_3$ and X-ray diffractogram after reduction of S–S bonds.[71] Reprinted with permission from *Chem. Commun.*, An original synthesis of highly ordered organosilica with high content of thiol groups by J. Alauzun, A. Mehdi, C. Reye, R.J.P. Corriu, 347. Copyright (2006) RSC

previously used, lamellar systems having a S–S bridge can be obtained.[71] Figure 5.42 shows the sequence of reactions leading to a lamellar material when $n = 11$. This material is then transformed into a functional material by reduction of the S–S bridge into two SH groups, followed by H_2O_2 oxidation of the SH groups into SO_3H.

The acid functions have been titrated. The SH functions have permitted the complexation of mercuric salts in the proportion of 1 Hg for 3.5 SH when $n = 3$ and 1 Hg for 2.6 SH when $n = 5$ (Figure 5.42).

Moreover, the chemistry of carbamates has permitted the preparation of another type of lamellar materials bearing NH_2 groups. This development is based on the ability of amine functions to react in CO_2 atmosphere to give ammonium carbamates in mild and reversible conditions [Equation (5.4)].

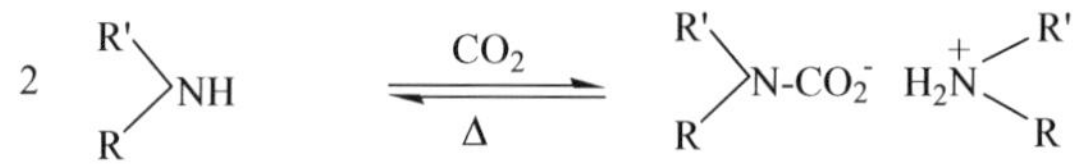

Equation 5.4

This reaction has been used to obtain lamellar materials bearing NH and NH_2 free groups (Figure 5.43).

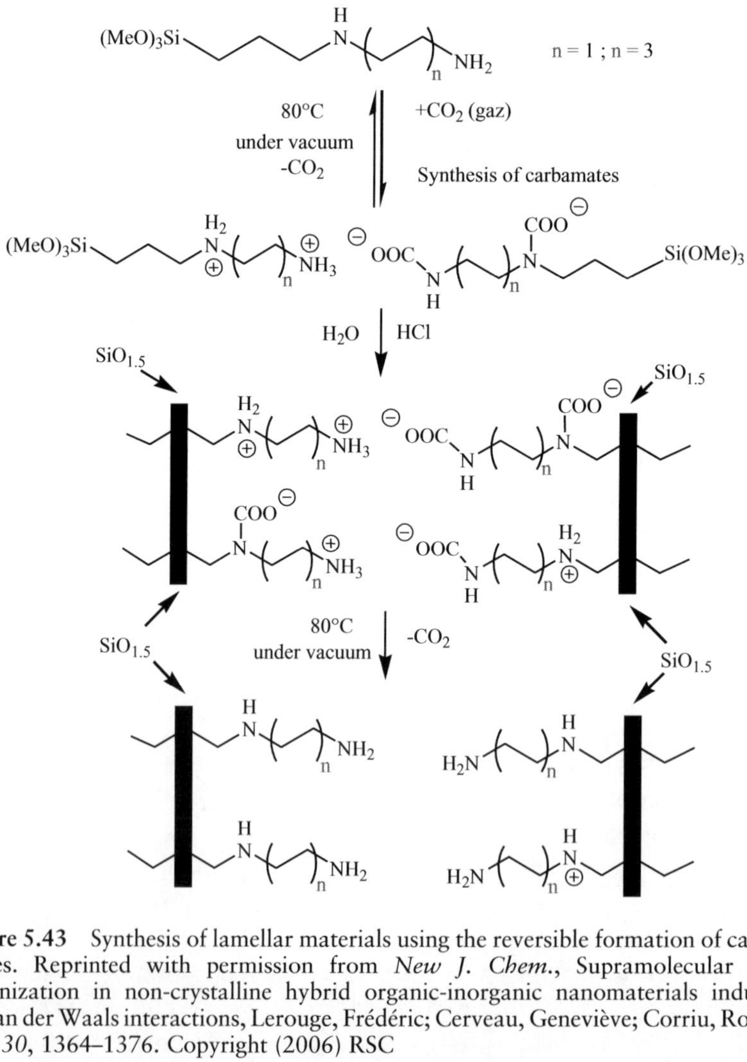

Figure 5.43 Synthesis of lamellar materials using the reversible formation of carbamates. Reprinted with permission from *New J. Chem.*, Supramolecular self-organization in non-crystalline hybrid organic-inorganic nanomaterials induced by van der Waals interactions, Lerouge, Frédéric; Cerveau, Geneviève; Corriu, Robert J.P., *30*, 1364–1376. Copyright (2006) RSC

Carbon dioxide reacts with commercially avalaible precursors such as $(MeO)_3Si(CH_2)_3NH(CH_2CH_2)_nNH_2$ ($n = 1$ or 3) to give a supramolecular network of ammonium carbamate. When this network is hydrolyzed in the presence of a stoichiometric amount of water at pH = 1.5, it leads to materials with a perfect lamellar structure when $n = 3$. Heating these materials at 80 °C under vacuum eliminates CO_2 and releases the amine functions without modification of the structure.[72]

These materials are excellent chelating reagents for transition metals and lanthanides. In some cases, it has been possible to chelate Eu^{3+} ions by only two nitrogen atoms. This unusual chelation can be explained by the fact that it does not occur in the liquid phase, but directly in the lamellar solid, which minimizes the entropy (*cf.* Section 5.8.4). The Eu^{3+} ions complete their sphere of coordination with the counter-ions (Cl^-, NO_3^-).

These results are promising for the development of materials capable of very efficiently chelating lanthanides and transition metals.

5.8 PROSPECTS

5.8.1 General Remarks

Although the chemistry of these nanostructured hybrid materials has been developed only recently, it opens up wide ranging and interesting possibilities. First of all, it is possible to incorporate and organize a large variety of chemical entities in a silica matrix. Until now, mostly molecules and metal complexes have been introduced into this type of material. However, nothing prevents the inclusion of more elaborate forms, such as metal particles, polymer fragments or inorganic combinations (cluster oxides, sulfides, phosphides, etc.). The only difficulty lies in the synthesis of precursors since the nano-objects that bring the properties must be substituted by several functionalities [generally $Si(OR_3)$] in order to allow their irreversible introduction into silica-like matrices.

We now briefly describe some possible (with our present knowledge) developments for this type of materials.

5.8.2 Properties Due to the Nano-Objects

The simplest materials have been obtained by inclusion of entities having physical properties. Two examples corresponding to photoluminescence and paramagnetism can be mentioned.

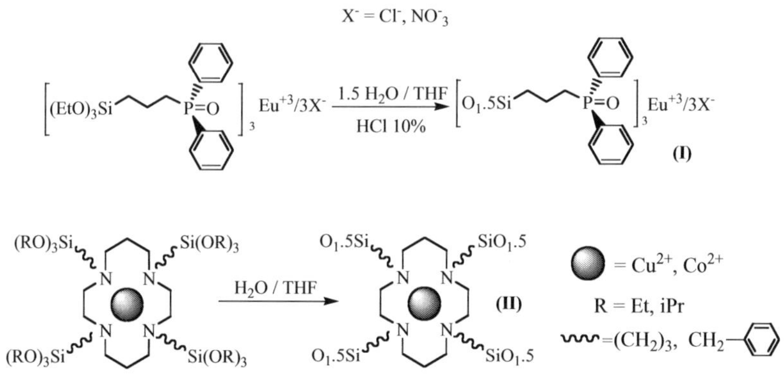

Figure 5.44 Photoluminescent (I) and paramagnetic (II) materials obtained by hydrolytic polycondensation of alkoxylated precursors

Two nano-objects have been prepared and polycondensed (Figure 5.44). The first one is a photoluminescent europium complex leading to material **I**[73] and the second is a paramagnetic Cu^{2+} complex giving material **II** after polycondensation.[74]

The photoluminescence of **I** is particularly intense. First of all, the Eu^{3+} ions are surrounded by six aromatic groups which produce an antenna effect enhancing the intensity. In addition, these aromatic groups provide a hydrophobic environment preventing coordination by H_2O molecules, which would strongly attenuate the photoluminescence. Finally, all the ions are in identical coordination sites.

In the other example, suitably substituted tetra-azamacrocycles permit the introduction of paramagnetic ions into the material. The paramagnetism observed is more intense than that of the precursor.

In conclusion, these single phase nanostructured materials are consequently highly reliable and capable of exploiting the nano-object's properties. The method allows the shaping of the material (fiber, matrix, film, etc.) and the efficient transformation of the nano-object into nanomaterial.

5.8.3 Influence of the Self-Organization on the Coordination Mode in the Solid

Study of the complexation of metal ions by nanostructured hybrid materials bearing free (*i.e.* non-complexed) tetra-azamacrocycle units has shown a behavior *a priori* surprising but easily rationalized.

The material can be prepared by two routes: polycondensation of the silylated complex (metal ion/cyclam) (route A); or polycondensation of

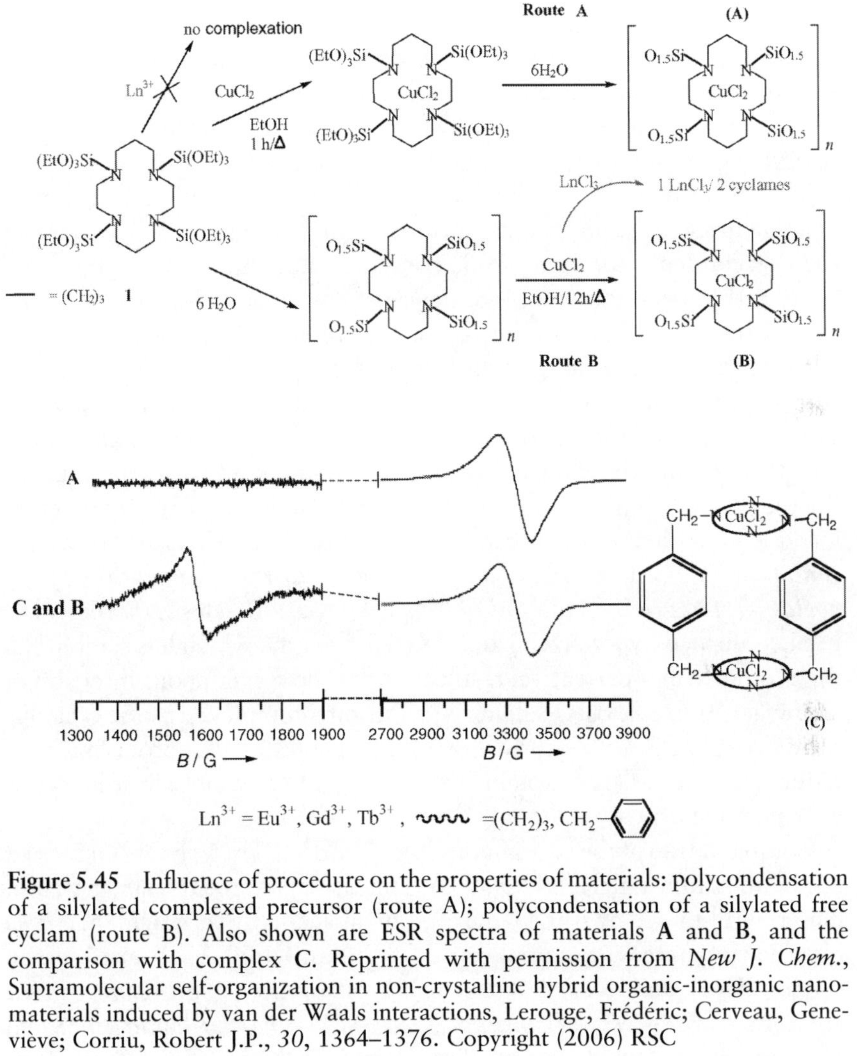

Figure 5.45 Influence of procedure on the properties of materials: polycondensation of a silylated complexed precursor (route A); polycondensation of a silylated free cyclam (route B). Also shown are ESR spectra of materials **A** and **B**, and the comparison with complex **C**. Reprinted with permission from *New J. Chem.*, Supramolecular self-organization in non-crystalline hybrid organic-inorganic nanomaterials induced by van der Waals interactions, Lerouge, Frédéric; Cerveau, Geneviève; Corriu, Robert J.P., *30*, 1364–1376. Copyright (2006) RSC

the 'free' silylated cyclam followed by complexation of the metallic salt on the resulting solid (route B). Figure 5.45 shows the two possible pathways as well as the EPR spectra of copper-containing materials, synthesized either by pathway A or B.[74]

The results obtained clearly show that materials **A** and **B** have very different characteristics. The ESR spectrum of **A** is in agreement with what is expected: it corresponds to an allowed transition at 3200 G ($\Delta M_s = 1$). In contrast, the spectrum of **B** is completely different. In addition to the allowed transition, a signal corresponding to a 'forbidden' transition[74] ($\Delta M_s = 2$) is observed; this is characteristic of Cu...Cu interactions. The same signal is observed in molecular complexes obtained from entities bearing two tetra-azamacrocycles bridged by a rigid spacer. In this case, the two Cu^{2+} ions lie face to face at a distance permitting the interaction (Figure 5.45, structure **C**).[75]

The phenomenon is explained by the fact that routes A and B cannot be equivalent since the solid precursors are totally different. Route A involves the polycondensation of an ionic complex presenting some rigidity and route B that of a non-complexed cyclam which can adopt various conformations. It is therefore not surprising that the self-organizations of these two entities are not identical. Indeed, the interactions between non-complexed cyclams and ion-complexed cyclams are obviously very dissimilar. The similarity of the ESR spectra of the molecule **C** and of the complex obtained by route B illustrates this point perfectly. In molecule **C**, the two Cu atoms are complexed by the cyclam units. They are therefore located at a definite distance from one another. *Consequently, this similarity of behavior suggests the presence of an organization of the chelating units in the material obtained by route B*. This hypothesis has been confirmed by a XANES and EXAFS[76*] study, which has established the geometry of a derivative, resulting from the spontaneous insertion of an oxygen molecule between the two Cu atoms complexed by the cyclams. In addition, the XANES study shows that one of the Cu^{2+} atoms has been reduced to Cu^{+}. (This explains the fixation of O_2 which, in solution, is only possible on Cu^{+}.)

Complexation of the lanthanides (Eu^{3+}, Gd^{3+}, Tb^{3+}) has been carried out on the same materials (**A** and **B**). The results obtained confirm that in materials obtained by route B the cyclam units are organized. In materials obtained by route A, no complexation of the lanthanide has been observed (Figure 5.45). Indeed, it is known that *in solution, the cyclam cannot complex lanthanides. In contrast, lanthanide ions can be incorporated into the solid obtained by route B with yields in the 45–47% range*. Thus, this solid has a chelating capacity unknown in solution, corresponding to the chelation of one lanthanide ion for two cyclams. This is explained by the self-organization of the cyclam units within the solid: lanthanide ions

---

* XANES and EXAFS techniques are based on X-ray absorption. They require a great intensity and are only possible with a particles accelerator. XANES measures the energy necessary for the abstraction of an electron from a given atom. EXAFS can determine whereat which distance are located the neighboring elements of the studied atom are located. From this method, a representation at least partial description of the local organization in an amorphous system can be obtained.

can be complexed because of the proximity of the nitrogen atoms of the two cyclams. In order to eliminate the possibility of ion adsorption by the SiOH present at the surface, the material has been copiously washed; however, the level of complexation remains the same after the washings. An EXAFS study carried out on this type of material has shown a statistical coordination of 7.5 to 9 nitrogen atoms surrounding a lanthanide (Eu^{3+}, Gd^{3+}, Tb^{3+}) ion. This technique demonstrates in addition that the Cl^- ions are not coordinated on the metal.

5.8.4 Coordination within the Solid: A New Experimentation Field

These results prove that a solid material bearing potentially chelating functions can be used as a chelating system *per se*, *thus opening the way to a coordination chemistry quite different from that in the liquid phase.* The interactions between the different elements are certainly the same. However, the solid has characteristics very different from those in solution. In a solid, the chelating entities are linked together and are located in a specific environment. *When the chelating units are organized, a decrease in entropy becomes possible:* the solid can organize the chelatants in ways that do not exist in solution. Thus, their relative arrangement in solid obtained by route B (Figure 5.45) can favor the coordination of ions, such as lanthanides, which may have coordination numbers higher than 6. The same type of chelation cannot be obtained in solution because of the very high entropy involved. *The observed self-organization in the solid can modify the ratio and the geometry of coordination of metallic ions.*

These observations open up new possibilities in the field of ion chelation. Indeed, coordination chemistry has been developed from reactions in solution. *Solids obtained by chelation in solution are thermodynamically controlled crystals.* A crystal thus obtained is thermodynamically stable since it results from a geometry which has 'survived' all possible exchanges in solution. In contrast, in a self-organized solid obtained by polycondensation of free chelating units, the coordination of ions can occur with a minimal entropy since the chelatants do not have to regroup around a metal center. It is, in contrast, the metal ion which is retained by the chelating units already organized in the solid. We can consider that it is *the organization of the material which will control the coordination of the metallic ions.* As an example, we can recall the coordination of a Eu^{3+} ion by two nitrogen atoms in the lamellar system described in Figure 5.43. Probably, the Eu^{3+} ions complete their coordination sphere with the counter-ions (Cl^-, NO_3^-).

This type of material is suitable for the study of new coordination complexes which do not exist in solution and whose coordination is not as well defined as in crystals. However, it is a true coordination which can resist repeated washings, and even some coordinating reagents. These complexes can *lead to new physical and chemical properties, since they are assemblages unknown until now.*

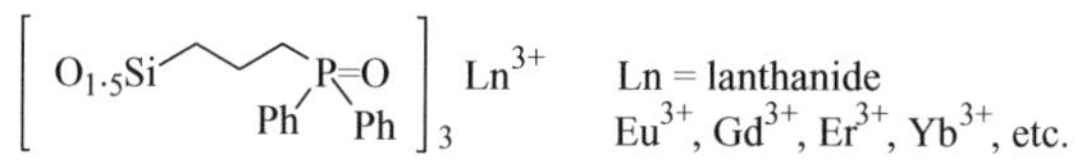

Figure 5.46 Examples of nanostructured materials with physical (optical, magnetic) properties

5.9 SOME POSSIBLE DEVELOPMENTS

Just to clarify the ideas, we shall now give some examples of possible developments from nanostructured systems.

5.9.1 Preparation of Nanomaterials from Nano-Objects

Nanostructured materials can be obtained from *any* nano-objects provided that they are functionalized by $-Si(OR)_3$ groups. Afterwards, it is only necessary to carry out the hydrolytic polycondensation of the nano-object and to shape the material into a film, matrix, fiber or coating.

According to the nature of the metallic salt, it will be possible to obtain materials having magnetic (Gd^{+3}, Fe^{+3}...) or optical (Eu^{+3}, Er^{+3}...) properties (Figure 5.46). The inclusion of metal nanoparticles or metallic oxides and the incorporation of molecules having physical (optical, nonlinear optical, conduction, magnetic, etc.) properties into the matrix are also conceivable (Figure 5.47). Only a limited number of studies have been performed in this field of research. However, this approach represents a very promising field, since it gives access to a new class of materials whose physical properties can be chosen and adjusted by the concentration of the nano-object.

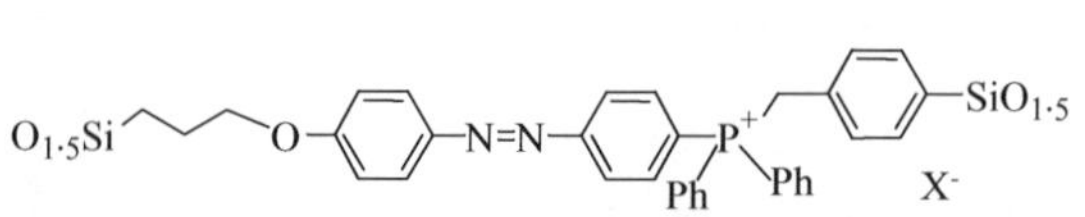

Figure 5.47 Nanostructured matrices obtained from a precursor having nonlinear optical properties

5.9.2 Nanostructured Hybrids as Matrices for Nanocomposite Material

Nanostructured hybrid materials show textural characteristics almost identical to those of SiO_2. It is even possible to adjust the dielectric constant. Therefore, nothing prevents their use as a matrix for shaping the material including the nano-object.

Nanoparticles of CdS[77,78] as well as metallic ions[79–81] have been introduced in a hybrid material having phenyl groups. No interactivity was expected. However, a regular distribution of CdS particles has been observed.

Clearly, the replacement of SiO_2 matrices by nanostructured hybrids opens up promising opportunities because of the possible interactions, at the nanometric scale, between the properties induced by the organic part of the matrix and those of the nano-object. These properties can be quite diverse (optical/magnetic, optical/electric, etc.). Their choice depends only on the intention and the expertise of the experimentalist. It is clear, however, that this is an untried domain, rich in possibilities but it can also yield some surprises. *Until now, this type of work has not been undertaken.*

5.9.3 Inclusion of Hybrid Systems in Matrices other than SiO_2

In Section 5.9.2, we mentioned the possibility of interactions between the organic part of the hybrid matrix and the nano-object. In this section, we study the possibility of interactions between the inorganic part – different from SiO_2 – and the organic part of a hybrid material. This is an important challenge. We have stressed (Chapters 3 and 4) the interest of matrices different from SiO_2. Indeed, SiO_2 has many qualities (stability, versatility and compatibility) with a great variety of (organic, inorganic, biological, etc.) entities. However, its only usable property is its adjustable dielectric constant. It is obvious that matrices capable of having other physical (optical, magnetic, electric, etc.) properties can lead to materials unavailable until now. They could allow the coupling of several properties at the nanometric scale: giving access to an enlarged 'nanoworld', so to speak. Let us recall that TiO_2 has interesting photovoltaic properties, SnO_2 is a semiconductor with a smaller gap than silicon, and NiO, Fe_3O_4, Fe_2O_3, etc. can lead to magnetic matrices. The number of possibilities is extensive.

However, as already mentioned, the hydrolysis of alkoxides and halides usually leads to precipitates. It is therefore necessary to resort to other precursors to obtain an adjustable solid by inorganic polymerization (sol-gel). The basics of this chemistry are known, since the coordination chemistry of all these elements has, at least in part, been studied. It should constitute the 'mine' that chemists wanting to obtain new matrices by inorganic polymerization should exploit to get adjustable solids. Preliminary studies have shown interesting possibilities for oxides, such as TiO_2 or SnO_2. Note, however, that the preparations of these two oxides are very different. Hydrolytic polymerization of TiO_2[82] has been carried out by precursors of the type $Ti(TEA)_2(OiPr)_2$ (where TEA stands for triethanolamine). As for SnO_2, hydrolysis of Sn-C bonds gives nanoporous SnO_2.[83] Materials with alternate layers of SnO_2/hydrophobic material have also been synthesized.[84–86] In addition, the existence of physical properties of these oxides requires crystalline organization. It will be necessary to obtain at least a microcrystalline powder, if not a unique crystal.

The development of inorganic polymerization, which permits access to oxide matrices by the sol-gel process, is a crucial challenge. For each oxide, *it will be important to choose the molecular precursors* among the set of available molecules. It will be necessary to test, by varying the experimental conditions, the precursors which will give rise to a process analogous to those existing for SiO_2. This chemistry should see important developments in the near future.

5.9.4 Functionalization of the Matrices

Clearly, using matrices having physical properties (TiO_2, SnO_2, Fe_2O_3. etc.) for studying the coupling of properties can only be done under some conditions. In particular, it will be necessary to perfect chemical methods allowing the incorporation of entities having physical properties into these matrices. Functionalization of oxides such as TiO_2, V_2O_5 or Fe_2O_3 cannot be done by direct grafting because of the lability of the metal–carbon bond in the presence of H_2O or OH groups. Grafting functions should be used to achieve this purpose. Only Si and P and, to a lesser degree, Sn and B can be used to bond covalently a function or an organic entity to a matrix, such as TiO_2, ZrO_2, V_2O_5 and Fe_3O_4.

Organic entities have been grafted on oxides by the intermediary $RSi(OR')_3$ or $RPO_3R''_2$ groups. The nature of the organic entity can be quite diverse (aryl, alkyl, etc.). $Si(OR')_3$ groups give satisfactory results for TiO_2, ZrO_2 and SnO_2. However, phosphonates can give thermodynamically stable assemblages with the majority of oxides.[87]

It is interesting to note that particles of different compositions (SiO_2, ZrO_2, etc.) have been assembled using bifunctional spacers:[88–96]

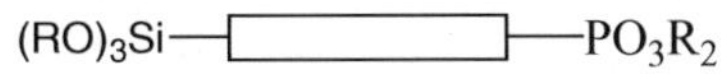

REFERENCES

[1] F.A. Cotton, G. Wilkinson, *Advanced Inorganic Chemistry*, John Wiley & Sons, Ltd, NewYork, **1980**.

[2] L. Haiduc, J.J. Zuckerman, *Basic Organometallic Chemistry*, W. de Gruyter, Berlin, **1985**.

[3] E.W. Colvin, *Silicon in Organic Synthesis*, Butterworth, London, **1981**.

[4] M.A. Brook, *Silicon in Organic, Organometallic, and Polymer Chemistry*, John Wiley & Sons, Ltd, NewYork, **2000**.

[5] B. Boury, R.J.P. Corriu, *The Chemistry of Organosilicon Compounds*, Z. Rappoport, Y. Apeloig,(eds), John Wiley & Sons, Ltd, Chichester, **2001**, *565*.

[6] B. Boury, R.J.P. Corriu, *Organosilicon Chemistry* V: From Molecules to Materials, Wiley-VCH, Weinheim, **2003**, 389.

[7] R. Corriu, *J. Organomet. Chem.* **2003**, *686*, 32.

[8] K.J. Shea, J. Moreau, D.A. Loy, R.J.P. Corriu, B. Boury, *Functional Hybrid Materials*, P. Gomez-Romero C. Sanchez,(eds), Wiley-VCH, Weinheim, **2004**, 50.

[9] B. Boury, R.J.P. Corriu, *Chem. Commun.* **2002**, 795.

[10] B. Boury, R.J.P. Corriu, *Chem. Record.* **2003**, *3*, 120.

[11] I. Ojima, *The Chemistry of Organic Silicon Compounds*, S. Pataï, Z. Rappoport,(eds), John Wiley & Sons, Ltd, Chichester, **1989**, 1479.

[12] R. Calas, J. Dunoguès, *J. Organomet. Chem. Library.* **1976**, *2*, 277.

[13] K.J. Shea, D.A. Loy, O. Webster, *J. Am. Chem. Soc.* **1992**, *114*, 6700.

[14] D.J. Brondani, R.J.P. Corriu, S. El Ayoubi, J. Moreau, M. Wong Chi Man, *J. Organomet. Chem.* **1993**, *451*, C1.

[15] D.J. Brondani, R.J.P. Corriu, S. El Ayoubi, J. Moreau, M. Wong Chi Man, *Tetrahedron Lett.* **1993**, *34*, 2111.

[16] D.J. Brondani, F.H. Carré, R.J.P. Corriu, J. Moreau, M. Wong Chi Man, *Angew. Chem., Int. Ed. Engl.* **1996**, *35*, 324.

[17] J.-P. Bezombes, C. Chuit, R.J.P. Corriu, C. Reye, *J. Mater. Chem.* **1998**, *8*, 1749.

[18] R.A. Benkeser, J.M. Gaul, W.E. Smith, *J. Am. Chem. Soc.* **1969**, *91*, 3666.

[19] R.F. Heck, *Org. React.* **1982**, *27*, 345.

[20] G. Cerveau, R.J.P. Corriu, E. Framery, *Chem. Mater.* **2001**, *13*, 3373.

[21] G. Cerveau, R.J.P. Corriu, C. Lepeytre, P.H. Mutin, *J. Mater. Chem.* **1998**, *8*, 2707.

[22] H.W., J. Oviatt, K.J. Shea, J.H. Small, *Chem. Mater.* **1993**, *5*, 943.

[23] M. Schneider, J. Hagen, D. Haarer, K. Mullen, *Adv. Mater.* **2000**, *12*, 351.

[24] B. Boury, R.J.P. Corriu, R. Numez, *Chem. Mater.* **1998**, *10*, 1795.

[25] P. Battioni, E. Cardin, M. Louloudi, B. Schöllhorn, G.A. Spyrioulas, D. Mansui, T.G. Traylor, *Chem. Commun.* **1996**, 2037.

[26] P. Chevalier, R.J.P. Corriu, P. Delord, J.J.E. Moreau, M. Wong Chi Man, *New J. Chem.* **1998**, *22*, 423.

[27] P. Chevalier, R.J.P. Corriu, J.J.E. Moreau, M. Wong Chi Man *J. Sol-Gel Sci. Technol.* **1997**, *8*, 603.
[28] R.J.P. Corriu, J.J.E. Moreau, P. Thepot, M. Wong Chi Man, C. Chorro, J.-P. Lere-Porte, J.-L. Sauvajol, *Chem. Mater.* **1994**, *6*, 640.
[29] R.J.P. Corriu, J.J.E. Moreau, P. Thepot, C. Chorro, J.-P. Lere-Porte, J. Sauvajol,-L. M. Wong Chi Man, *Synth. Met.* **1994**, *62*, 233.
[30] R.J.P. Corriu, J.J.E. Moreau, P. Thepot, M. Wong Chi Man, *Chem. Mater.* **1996**, *8*, 100.
[31] R.J.P. Corriu, A. Mehdi, C. Reye, C. Thieuleux, *Chem. Mater.* **2004**, *16*, 159.
[32] R.J.P. Corriu, A. Mehdi, C. Reye, C. Thieuleux, *New J. Chem.* **2003**, *27*, 905.
[33] G. Cerveau, R.J.P. Corriu, E. Framery, *Chem. Commun.* **1999**, 2081.
[34] G. Cerveau, R.J.P. Corriu, E. Framery, *J. Mater. Chem.* **2000**, *10*, 1617.
[35] G. Cerveau, R.J.P. Corriu, E. Framery, *J. Mater. Chem.* **2001**, *11*, 713.
[36] G. Cerveau, R.J.P. Corriu, E. Framery, S. Ghosh, P.H. Mutin, *J. Mater. Chem.* **2002**, *12*, 3021.
[37] G. Cerveau, R.J.P. Corriu, C. Fischmeister-Lepeytre, *J. Mater. Chem.* **1999**, *9*, 1149.
[38] G. Cerveau, R.J.P. Corriu, E. Framery, *Polyhedron.* **2000**, *19*, 307.
[39] G. Cerveau, R.J.P. Corriu, E. Framery, *C. R. Acad. Sci.* **2001**, *4*, 79.
[40] J.J.E. Moreau, L. Vellutini, M. Wong Chi Man, C. Bied, J.-L. Bantignies, P. Dieudonne, J.-L. Sauvajol, *J. Am. Chem. Soc.* **2001**, *123*, 7957.
[41] J.J.E. Moreau, L. Vellutini, M. Wong Chi Man, C. Bied, *J. Am. Chem. Soc. 123*, **2001**, 1509.
[42] J.J.E. Moreau, L. Vellutini, M. Wong Chi Man C. Bied, *Chem. -Eur. J.* **2003**, *9*, 1594.
[43] C. Bied, J.J.E. Moreau, L. Vellutini, M. Wong Chi Man, *J. Sol-Gel Sci. Technol.* **2003**, *26*, 583.
[44] J.J.E. Moreau, B.P. Pichon, M. Wong Chi Man, C. Bied, H. Pritzkow, J.-L. Bantignies, P. Dieudonne, J.-L. Sauvajol, *Angew. Chem., Int. Ed.* **2004**, *43*, 203.
[45] J.J.E. Moreau, B.P. Pichon, C. Bied, M. Wong Chi Man, *J. Mater. Chem.* **2005**, *15*, 3929.
[46] J.J.E. Moreau, L. Vellutini, P. Dieudonne, M. Wong Chi Man, J.-L. Bantignies, J.-L. Sauvajol, C. Bied, *J. Mater. Chem.* **2005**, *15*, 4943.
[47] G. Cerveau, R.J.P. Corriu, C. Lepeytre, *J. Mater. Chem.* **1995**, *5*, 793.
[48] G. Cerveau, R.J.P. Corriu, C. Lepeytre, *Chem. Mater.* **1997**, *9*, 2561.
[49] G. Cerveau, R.J.P. Corriu, J. Dabosi, J.L. Aubagnac, R. Combarieu, Y. de Puydt, *J. Mater. Chem.* **1998**, *8*, 1761.
[50] G. Cerveau, R.J.P. Corriu, J. Dabosi, C. Fischmeister-Lepeytre, R. Combarieu, *Rapid. Commun. Mass Spectrom.* **1999**, *13*, 2183.
[51] B. Boury, R.J.P. Corriu, P. Delord, M. Nobili, V. Le Strat, *Angew. Chem., Int. Ed.* **1999**, *38*, 3172.
[52] F. Ben, B. Boury, R.J.P. Corriu, V. Le Strat, *Chem. Mater.* **2000**, *12*, 3249.
[53] B. Boury, F. Ben, R.J.P. Corriu, M. Nobili, P. Delord, *Chem. Mater.* **2002**, *14*, 730.
[54] A. Guignier, *Théorie de la Radiocristallographie*, Dunod, Paris, **1955**.
[55] B.K. Vainsthein, *Diffraction of X-Rays by Chain Molecules*, Elsevier, New York, **1966**.
[56] G. Cerveau, R.J.P. Corriu, E. Framery, F. Lerouge, *Chem. Mater.* **2004**, *16*, 3794.
[57] G. Cerveau, R.J.P. Corriu, E. Framery, F. Lerouge, *J. Mater. Chem.* **2004**, *14*, 3019.
[58] B. Boury, F. Ben, R.J.P. Corriu, *Adv. Mater.* **2002**, *14*, 101.

[59] F. Lerouge, G. Cerveau, R.J.P. Corriu, Unpublished results.
[60] P.J. Collings, M. Hird, *Introduction to Liquid Crystals: Chemistry and Physics*, Taylor & Francis, London, **1997**.
[61] F. Hensel, *Z. Phys. Chem. (Munich)* **2000**, *214*(11), 1622.
[62] F. Lerouge, G. Cerveau, R.J.P. Corriu, *J. Mater. Chem.* **2006**, *16*, 90.
[63] A. Pegenau, T. Hegmann, C. Tschierske, S. Diele, *Chem. -Eur. J.* **1999**, *5*, 1643.
[64] H. Helker, R. Hatz, *Handbook of Liquid Crystals*, Verlag Chemie, Weinheim, **1980**.
[65] G. Cerveau, R.J.P. Corriu, F. Lerouge, N. Bellec, D. Lorcy, M. Nobili, *Chem. Commun.* **2004**, 396.
[66] N. Bellec, F. Lerouge, B. Pichon, G. Cerveau, R.J.P. Corriu, D. Lorcy, *Eur. J. Org. Chem.* **2005**, 136.
[67] N. Bellec, F. Lerouge, O. Jeannin, G. Cerveau, R.J.P. Corriu, D. Lorcy, *J. Organomet. Chem.* **2006**, *691*, 5774.
[68] F. Lerouge, G. Cerveau, R.J.P. Corriu, C. Stern, R. Guilard, *Chem. Commun.* **2007**, 1553.
[69] P. Grosshanns, A. Jouaiti, V. Bulach, J.-M. Planeix, M.W. Hosseini, J.-F. Nicoud, *Chem. Commun.* **2003**, 1336.
[70] J. Alauzun, A. Mehdi, C. Reye, R.J.P. Corriu, *J. Mater. Chem.* **2005**, *15*, 841.
[71] J. Alauzun, A. Mehdi, C. Reye, R.J.P. Corriu, *Chem. Commun.* **2006**, 347.
[72] J. Alauzun, A. Mehdi, C. Reye, R.J.P. Corriu, *J. Am. Chem. Soc.* **2005**, *127*, 11204.
[73] F. Embert, A. Mehdi, C. Reye, R.J.P. Corriu, *Chem. Mater.* **2001**, *13*, 4542.
[74] G. Dubois, C. Reye, R.J.P. Corriu, S. Brandes, F. Denat, R. Guilard, *Angew. Chem., Int. Ed.* **2001**, *40*, 1087.
[75] M. Lachkar, R. Guilard, A. Atmani, A. De Cian J. Fisher, R. Weiss, *Inorg. Chem.* **1998**, *37*, 1575.
[76] J. Goulon, C. Goulon-Ginet, A. Rogalev, F. Wilhelm, N. Jaouen, D. Cabaret, Y. Joly, G. Dubois, R.J.P. Corriu, G. David, S. Brandes, R. Guilard, *Eur. J. Inorg. Chem.* **2005**, 2714.
[77] K.M. Choi, K.J. Shea, *Chem. Mater.* **1993**, *5*, 1067.
[78] K.M. Choi, K.J. Shea, *J. Phys. Chem.* **1994**, *98*, 3207.
[79] K.M. Choi, K.J. Shea, *J. Am. Chem. Soc.* **1994**, *116*, 9052.
[80] K.M. Choi, K.J. Shea, *Mater. Res. Soc. Symp. Proc.* **1994**, *346*, 763.
[81] K.M. Choi, K.J. Shea, *J. Sol-Gel Sci. Technol.* **1995**, *5*, 143.
[82] R.J.P. Corriu, A. Mehdi, C. Reye, Y. Guari, G.X. Feng,Unpublished results.
[83] T. Toupance, H. El Hamzaoui, B. Jousseaume, H. Riague, I. Saadeddin, G. Campet, J. Broetz, *Chem. Mater.* **2006**, *18*, 6364.
[84] H. Elhamzaoui, B. Jousseaume, H. Riague, T. Toupance, P. Dieudonne, C. Zakri, M. Maugey, H. Allouchi, *J. Am. Chem. Soc.* **2004**, *126*, 8130.
[85] G. Vilaca, B. Jousseaume, C. Mahieux, C. Belin, H. Cachet, M.-C. Bernard, V. Vivier, T. Toupance, *Adv. Mater.* **2006**, *18*, 1073.
[86] H. Elhamzaoui, B. Jousseaume, T. Toupance, C. Zakri, M. Biesemans, R. Willem, H. Allouchi, *Chem. Commun.* **2006**, 1304.
[87] A. Vioux, J. Le Bideau P.H. Mutin, D. Leclercq, *Top. Curr. Chem.* **2004**, *232*, 145.
[88] C. Carbonneau, R. Frantz, J.-O. Durand, G.F. Lanneau, R.J.P. Corriu, *Tetrahedron Lett.* **1999**, *40*, 5855.
[89] C. Carbonneau, R. Frantz, J.-O. Durand, M. Granier, G.F. Lanneau, R.J.P. Corriu, *New J. Chem.* **2001**, *25*, 1398.
[90] C. Carbonneau, R. Frantz, J.-O. Durand, M. Granier, G.F. Lanneau, R.J.P. Corriu, *J. Mater. Chem.* **2002**, *12*, 540.

[91] R. Frantz, C. Carbonneau, M. Granier, J.-O. Durand, G.F. Lanneau, R.J.P. Corriu, *Tetrahedron Lett.* **2002**, *43*, 6569.
[92] R. Frantz, J.-O. Durand, G.F. Lanneau, J. Olivier-Fourcade, J.-C. Jumas, M. Crétin, M. Persin, *Eur. J. Inorg. Chem.* **2002**, 1088.
[93] R. Frantz, M. Granier, J.-O. Durand, G.F. Lanneau, *Tetrahedron Lett.* **2002**, *43*, 9115.
[94] P. Audebert, S. Sadki, F. Miomandre, G.F. Lanneau, R. Frantz, J.-O. Durand, *J. Mater. Chem.* **2002**, *12*, 1099.
[95] C. Carbonneau, J.-O. Durand, M. Granier, G.F. Lanneau, *Chem. Commun.* **2003**, 1166.
[96] R. Frantz, J.-O. Durand, M. Granier, G.F. Lanneau, *Tetrahedron Lett.* **2004**, *45*, 2935.

6

Chemistry Leading to Interactive Nanomaterials

6.1 INTRODUCTION

Interactive nanomaterials are the nanometric version of 'smart materials'. The latter, prepared by classical methods, were and still are the subject of many interesting investigations. Smart materials couple several properties of the same material. Interactive nanomaterials also aim to couple several (physical, chemical, mechanical, etc.) properties, but this coupling is interactively controlled at the nanometric scale.

Interactive nanomaterials can be considered as materials, which will be developed in the near future. Therefore, it will not be possible to describe them in detail, since their very existence is now limited to some preliminary investigations. *Most of the chemistry for their preparation remains to be discovered.* However, existing data and knowledge can guide the study of this chemistry. *Note that until now the concept of the interactive coupling of properties was totally unknown to molecular chemistry.* Molecular chemistry has never been thought of in terms of localization (see Section 6.6.2) and, even more so, in terms of interactive localization, since it originates from liquid phase methodologies.

The design of interactive nanomaterials is entirely new, so we shall limit ourselves to an account of the research routes and ways of thinking, which could help chemists to create combinations of properties whose interactions may be controlled. Emphasis will be put on the preparative methods of this type of materials.

Molecular Chemistry of Sol-Gel Derived Nanomaterials Robert Corriu and Nguyên Trong Anh

Prior to this discussion, it is useful, however, to recap the current knowledge on interactive properties in material science.

6.2 SMART MATERIALS

Materials capable of coupling several properties already exist, independently of nanosciences. The newest offspring of material science, they are solids presenting at least two coupled properties. Let us outline the subject.

The objective is the preparation of materials having two functions: generally a detection function coupled with a reaction function. The material must be able to spontaneously modify its physical (form, color, elasticity, etc.) properties in response to external factors (variations of temperature, mechanical constraints, electrical or magnetic field, etc.). It must be possible for the material to perceive, even to signal, a modification occurring in its environment. Moreover, it can adapt a reaction in order to induce a corrective effect. This is an exceedingly exciting research domain, which has already experienced some success. For instance, it is possible to detect structural weaknesses in an aeroplane framework or cracks in a building. One can also insert filters in arteries, which will deploy to reduce the risk of dispersion of blood clots.

What are the most advanced smart materials at present? They are principally alloys having shape memory (SMA) and piezoelectric (electrostrictive or magnetostrictive) materials.

Distorted when cold, the SMAs recover their initial form beyond a certain temperature. This is due to a temperature-dependent reversible transformation of the material's shape, corresponding to a phase transition in the solid. This property is observed in nickel-titanium alloys (Nitinol) modified by addition of other elements (Cu, Fe, Cr). Since the late 1960s, the weapons and electronic industries have used these alloys in hydraulic pipes or electric collectors. General everyday products exploiting these properties include thermostats, carburetors, toys and sculptures.

The most spectacular example involves filters with shape memory capable of trapping blood clots in vessels. A tiny filament of alloy with shape memory is cooled under room temperature. It can easily enter a catheter and be placed without surgery into a large blood vessel. Once warmed to room temperature, it deploys and becomes a filter, firmly maintained in place. It can thus recover the blood clots circulating in the blood and prevents them entering small vessels of essential organs (brain, heart or lungs).

Systems using piezoelectric properties can monitor civil engineering works (bridges, dams, etc.). Indeed, concrete constructions suffer stresses due to various causes (weight, vibrations, frost, earthquakes, etc.). Consequently, they can be damaged suddenly without their defects being spotted by preventive inspections. Scientists from the University of New York at Buffalo (D. Chung) had the good idea of introducing piezoelectric carbon fibers at the most sensitive sites, thus creating a proper 'nervous system' within the concrete. The fibers less than 10 μm in diameter and only centimeters long are mixed with the concrete during its preparation. They stick out of the surface and give a good electrical contact. By measuring the electrical resistance of the fiber, it is possible to detect a stress change.

The so-called 'intelligent glasses' having self-correction properties can also be mentioned. The most common application is a surface treatment with TiO_2 to give self-cleaning windows. The grime deposited on the surface can be eliminated by oxidation under sunlight.

More sophisticated applications exist: for example, a deposit of liquid crystals can modify the transparency of windows, which automatically darken under a bright sun and regulate the degree of transparency according to the luminosity level.

Research on this type of materials is very active and opens up very exciting prospects. *However, this field does not belong to nanosciences but rather to material science.*

6.3 THE ROUTE TO INTERACTIVE MATERIALS – DEFINITIONS

In this section, we shall discuss the state-of-the-art chemical routes to interactive nanomaterials. This chemistry is still in its infancy. Nonetheless, nanomaterials capable of coupling several properties at the nanometric scale have been prepared. They have been obtained by inorganic polymerization and are thus kinetically controlled nanomaterials.

The first materials were made in 1992, following a modification of the standard procedure. Scientists of the Mobil Company working on zeoliths had the idea to try a micellar medium in place of the usual quaternary ammonium templates. By doing this, Dr Kresge's group turned a page in the history of chemistry and started a new chapter with the discovery of mesoporous materials.*[1,2] It must be noted, however, that this success had

* Strictly speaking these materials are nanoporous. Discovered in 1972, before the formulation of the nanosciences concept, they have been called mesoporous materials. We shall keep this terminology.

been preceded by preliminary works producing lamellar silicates by intercalation of quaternary ammonium.[3]

6.4 MESOPOROUS MATERIALS

6.4.1 Production

Mesoporous materials are obtained in one step by hydrolytic polycondensation (sol-gel process) of a metal alkoxide [$Si(OEt)_4$in most cases]. The reaction is carried out in water with a surfactant in micellar concentration. Polycondensation occurs around the micelle without destroying it. However, the geometry of the micelle and the course of the reaction are heavily modified. The materials obtained after elimination of the surfactant are porous. The pores are of even size, generally with hexagonal-type organization. Specific surfaces are very high (between $500\,m^2\,g^{-1}$ and $1500\,m^2\,g^{-1}$) (Figure 6.1).

In the early experiments, the surfactants used were ionic (e.g. cetyltrimethylammonium bromide) and their elimination achieved by calcination. However, calcination and organic functions are incompatible and new methods were developed, taking advantage of the large variety of existing surfactants.[4–8] Improved synthetic methods can now control the size of the pores and, more interestingly, their functionalization (see below).

6.4.2 Some Examples of Mesoporous Silica

To illustrate the versatility of the method, we shall first describe succinctly the preparation of mesoporous silica. Only a limited number of examples will be presented because the literature on this subject has undergone an exponential growth. Recent advances are reported in several reviews.[9,10]

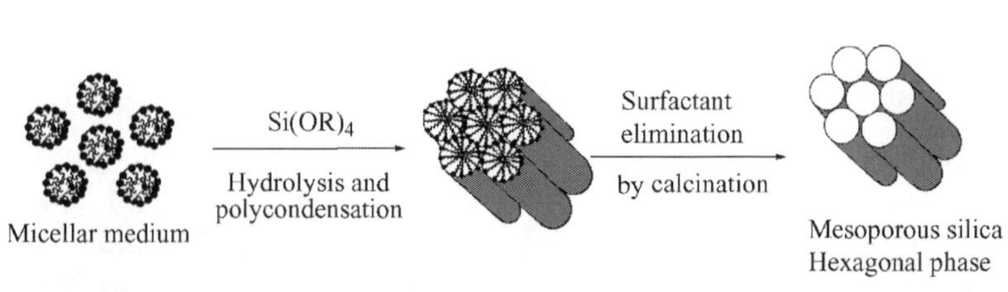

Figure 6.1 Hexagonal micellar organization in the colloidal sol phase. Formation of a porous silica presenting a regular pore distribution of hexagonal type after removal of the surfactant. Reprinted with permission from *Journal of Materials Chemistry*, Molecular chemistry and nanosciences: on the way to interactive materials by Robert Corriu, Ahmad Mehdi, Catherine Reye, *15*, 1. Copyright (2005) RSC

Later, we shall expound the prospects of these materials, from the point of view of nanomaterial development.

It can be stated that, at present, perfectly organized mesoporous silica can be found in everyday objects. These materials present a hexagonal structure of pores, whose sizes range between 3 nm and 20 nm. Other phases, cubic or lamellar, have been obtained but little studied. Consequently, we shall not discuss them, since the aim of this book is the presentation of the most useful tools for nanosciences. From this point of view, the most promising materials correspond to the hexagonal phases, which have furnished stable, well-organized materials, with large porous surfaces and a very stable framework. They also permit the introduction of a large spectrum of functionalities. Finally, these materials can be shaped by spin or dip coating to give films of calibrated thickness. These techniques are the most frequently used methods for transforming a sol into materials in thin layers, whose thickness, between hundreds of nanometers and tens of micrometers, can be experimentally controlled.[11–14]

Different types of materials which can be shaped and which present stable and functionalized structures have been described. Table 6.1 shows the principal types of mesoporous silica. This list is not exhaustive since

Table 6.1 The principal types of mesoporous silica

Type	Surfactant used	Structure	Pore diameter (nm)	Specific Surface m^2g^{-1}
MCM Mobil Corporation Materials	Cationic[a] CTAB CTAC	Hexagonal	3–20	1000–1400
HMS Hexagonal Mesoporous Solids	Non-ionic[a] NH_2	Wormlike	2–3	1000–1200
MSU Michigan State University	Non-ionic[b] diblock[c]	Hexagonal	2–4	700–1000
SBA Santa Barbara	Non-ionic[b] triblock[d] P123 F127	Hexagonal	5–10	600–800

[a]Cetyltrimethyl ammonium bromide (CTAB) chloride (CTAC).
[b]Docecylamine or hexadecylamine.
[c]$Brij_{56}$, $C_{16}H_{33}(PEO)_{10}$; $C_{13}H_{33}(PEO)_{12}$.
[d]P123, $(PEO)_{20}(PPO)_{70}(PEO)_{20}$; F127, $(PEO)_{110}(POE)_{70}(PEO)_{110}$.
PEO, poly(ethylene oxide); PPO, poly(propylene oxide).

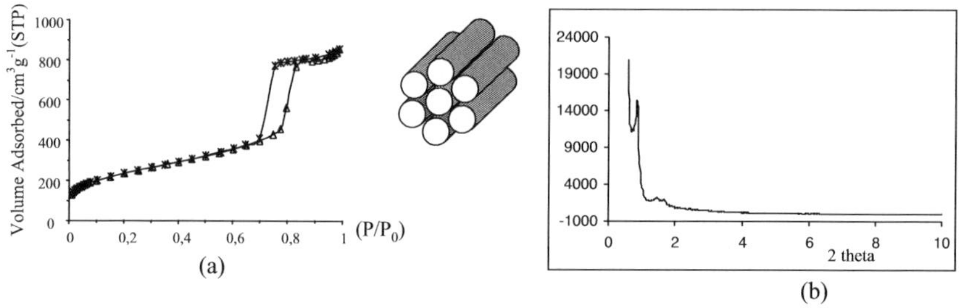

Figure 6.2 (a) Adsorption – desorption isotherm and (b) X-ray diffractogram of a mesoporous silica with a two-dimensional hexagonal structure. Reprinted with permission from *Journal of Materials Chemistry*, Corriu, Robert J. P.; Lancelle-Beltran, Emmanuelle; Mehdi, Ahmad; Reye, Catherine; Brandes, Stephane; Guilard, Roger. *12*, 1355–1362. Copyright (2002) RSC

many other surfactants have been used. Only the most frequently employed surfactants giving the best results are mentioned.

The hexagonal distribution of the pores is shown by X-ray diffractograms. The regularity of the nitrogen adsorption–desorption isotherms is a good indication that the pore diameters are monodisperse (Figure 6.2).

Note that the size of the pores is adjustable. As a matter of fact, it is possible to control the micelle's size by introducing variously substituted aromatic compounds (benzene, toluene, *p*-xylene, durene, mesitylene, etc.).[2,15]

More recently, the methodology has been improved by replacing $Si(OEt)_4$ and $Si(OMe)_4$ with $Si(OCH_2CH_2OH)_4$. The precursor $Si(OCH_2CH_2OH)_4$ is water-soluble and the reaction occurs without a catalyst.[16,17]

6.5 FUNCTIONALIZATION OF THE PORES

Two methods permit the introduction of chemical functions into the pores. The simplest way is a graft on suitable mesoporous silica. Another method, called 'direct synthesis', introduces the functional groups during the preparation of the mesoporous material. Let us examine and compare these two methods.

6.5.1 Functionalization by Grafting

Grafting functions on the internal surface of pores of any mesoporous material is accomplished using a molecule bearing a $Si(OR)_3$ group, a

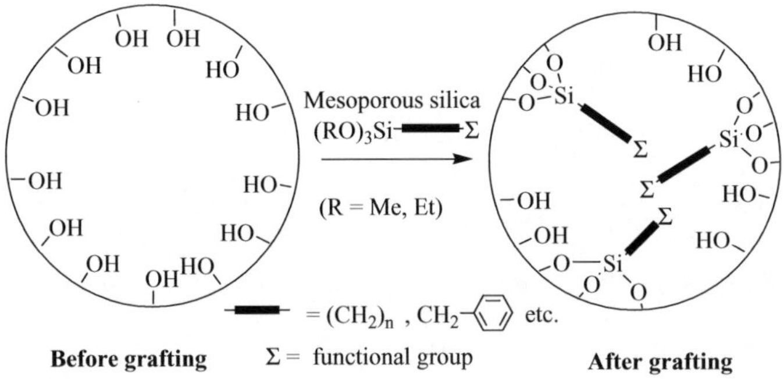

Figure 6.3 Outline of pore functionalization by grafting

chemical spacer and a function Σ [$\Sigma = NH_2$, SH, $P(C_6H_5)_2$, Cl, CN, etc.]. The $Si(OR)_3$ group, reacting with the SiOH groups on the pore surface, allows the functionalization of the material (Figure 6.3).[9,10,18–21]

The range of functions, which can be introduced, is very large. The method is general and permits in particular the introduction of voluminous groups without structure modification.

However, this grafting method, although very convenient, is not optimal for several reasons. It leads to a sizeable reduction in the porous volume and the surface of the pores. In addition, the distribution of the functions in the pores is irregular since it depends on several parameters, in particular on the distribution of the SiOH functions in the pores and on the size of the groups to be grafted. And, finally, the quantity of grafted functions is not controllable and is always inferior to that obtained by the direct method (see below).

We can consider grafting as a very convenient method for the insertion of a certain number of functions in the pores, when direct synthesis cannot be used. Note, in particular, the possibility of adjusting the hydrophobicity of the pores by grafting aliphatic functions, such as $(RO)_3Si\text{-}(CH_2)_n\text{-}(CH_3)$ (with $n = 1, 2, 3, 5$), or simply by a surface treatment with $(Me_3Si)_2NH$.

6.5.2 Functionalization by Direct Synthesis

This method introduces functional groups during the synthesis of the mesoporous material. The process consists of adding the molecule bearing the $(RO)_3Si\text{-}\Sigma$ chemical function directly into the reacting mixture [Si $(OR)_4$, catalyst, H_2O, surfactant]. This method is based on the capacity of micelles to incorporate hydrophobic species (Figure 6.4).[22–28]

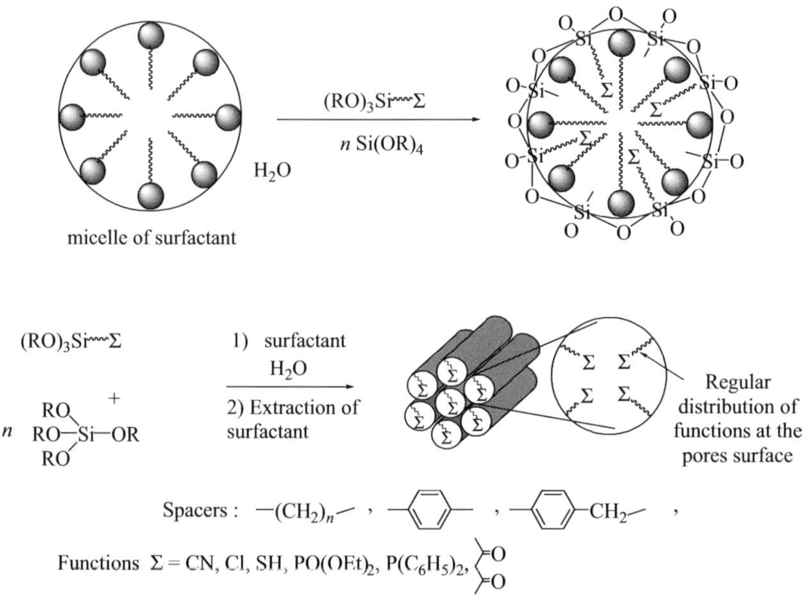

Figure 6.4 Outline of pore functionalization by direct synthesis. Nature of the spacers and of the introduced functions

All molecules used in direct synthesis have a hydrophobic part capable of passing into the micelle and a $Si(OR)_3$ head. This head generates SiOH bonds during the hydrolysis and thus, becomes polar. Consequently, the function is placed inside the pores through the intermediary of the micelle. At the same time the SiOH groups are attached on the wall.

This method has several advantages. It is possible in one step to obtain silica, functionalized at the internal surface of the pores in a very regular manner. In addition, the quantity of organic functions introduced can be controlled; a quantity often significantly superior to that obtained by grafting. The maximum introduced corresponds to one function for four SiO_2 units in the material. This ratio depends on the nature of the functional group and on the surfactant used. It is clear that, as in any systems kinetically controlled, it is necessary to optimize the experimental conditions. Thus, introduction of a high proportion of functions results in a loss of material organization. This is due to the perturbation introduced in the micellar medium. Organized systems can only be obtained for (function/SiO_2) ratios smaller than 1/10. The range of functions that can be introduced is very large. Some examples are given in Figure 6.4. In

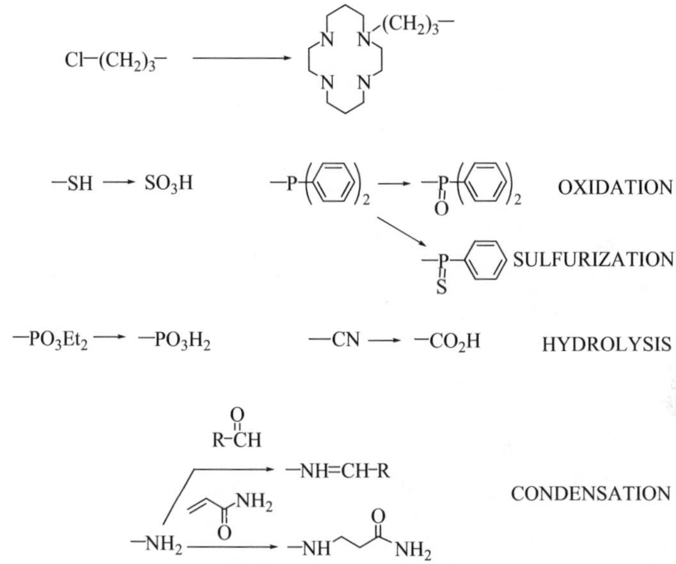

Figure 6.5 Some transformations carried out on functions located in the pores of mesoporous materials

addition, the functions located inside the pores can react quantitatively without modification of the material's structure (Figure 6.5).

It is important to stress that these materials are only usable if they are not alterable. The stability of the final material is ensured by a hydrothermal synthesis (100 °C/HCl). This treatment terminates the polycondensation reactions on silicon atoms, and further modification of the material is thus avoided.

The $-NH_2$ group deserves a special mention. This function is very useful, as it can undergo many organic reactions. In addition, amines are coordination agents for metal ions and particles, hence their applications in catalysis. They can also be transformed into polydentate chelatants, very efficient for selective separations. Finally, they can be used to attach biological molecules.

Direct synthesis of silica functionalized by amino groups is only possible in the presence of ionic surfactant (cetyltrimethylammonium bromide)[29,30] in basic medium or in the presence of non-ionic surfactant such as long chain amines (cetylamine, $C_{16}H_{23}NH_2$).[23] Unfortunately, both methods produce small size pores (<30 Å). Amines are also incompatible with surfactant of the Brij or Pluronic type (Table 6.1). The reason is that these surfactants are block copolymers bearing ether oxide units, generally used in acid medium to give a better organization. Now, in these

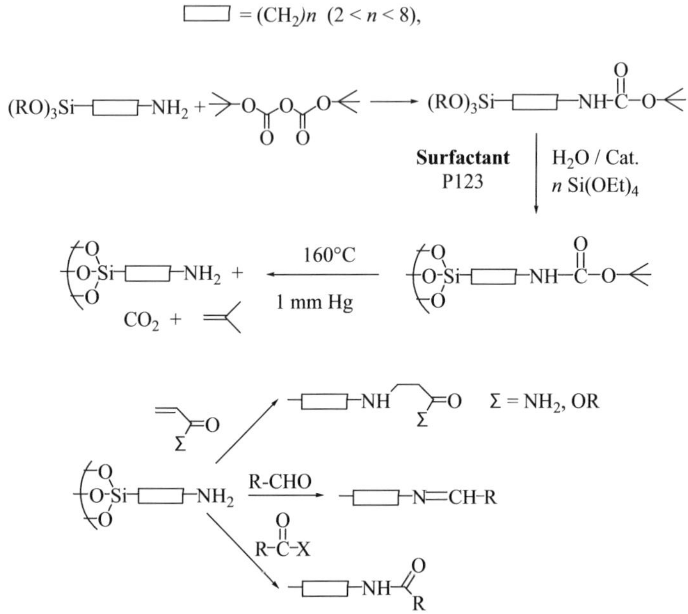

Figure 6.6 Outline of the protection by the Boc group followed by some functional transformations

conditions, the amine functions are protonated ($-NH_3^+$) and therefore located at the hydrophilic part and not in the hydrophobic micelle. The functions are then incorporated into the matrix and not in the pores. Furthermore, deprotonation implies the use of a strong base, which reacts with SiO_2 and may modify the material's structure.

The only method[31] which gives unequivocally $-NH_2$ groups located inside pores of > 30 Å requires protection of the amine function by the *tertio*-butyloxycarbonyl (Boc) group, commonly used as a protector group in peptide synthesis (*cf.* Chapter 4.1).

Figure 6.6 illustrates some possibilities offered by NH_2 groups for the synthesis of nanomaterials having chelating properties.

6.6 FUNCTIONALIZATION OF THE FRAMEWORK

The prospects opened up by mesoporous materials have aroused an increasing interest in their possibilities. Functionalization of the pores remains a domain of great importance. However, functionalization of the framework is opening up new and vast possibilities. This scientific quest started with the discovery of periodic mesoporous organosilica (PMO), which was made[32–34] when $Si(OR)_4$ was replaced by an organosilicon compound bearing several $Si(OR)_3$ groups.

6.6.1 Production of Periodic Mesoporous Organosilica

The first papers on the synthesis of PMOs have been published only recently. They describe the synthesis of evenly organized mesoporous systems in which the framework is not made of SiO_2 but of polysilylated organic units. The organization and functionalization of the pores remain similar to those in the silica case.

These nanomaterials have hexagonal structures.[32–39] Using aromatic groups, and after hydrothermal treatment, a quasicrystalline framework has been observed.[38]

However, the first organic units used [$(CH_2)_m$, –CH=CH–, thiophene, pheny]) led to materials without usable properties. More recent works have shown that it is possible to introduce in the framework functional organic entities capable of bringing in (physical, chemical or other) properties.[40–47] Polycondensation occurs in the presence of Si $(OR)_4$. The most significant results are shown in Figure 6.7.

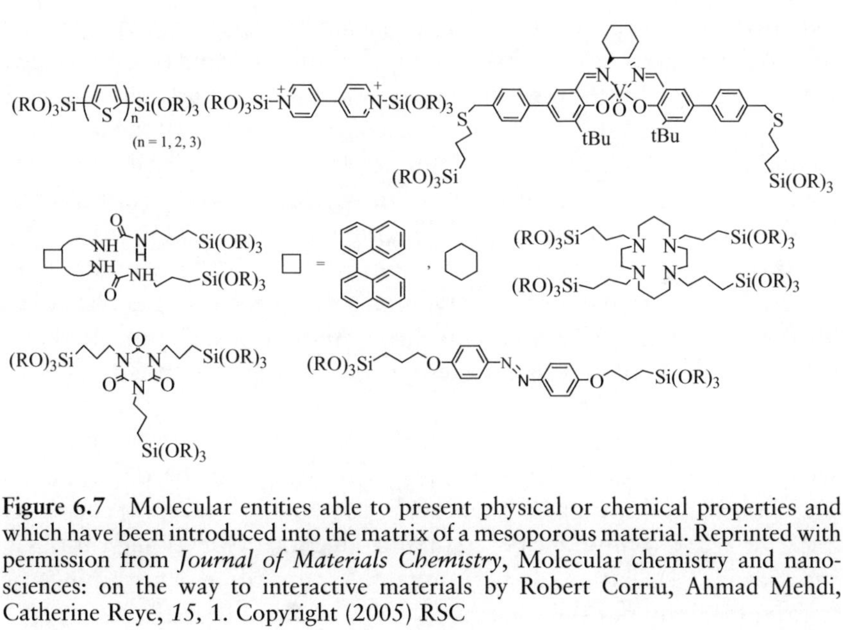

Figure 6.7 Molecular entities able to present physical or chemical properties and which have been introduced into the matrix of a mesoporous material. Reprinted with permission from *Journal of Materials Chemistry*, Molecular chemistry and nanosciences: on the way to interactive materials by Robert Corriu, Ahmad Mehdi, Catherine Reye, *15*, 1. Copyright (2005) RSC

These entities often occupy an important volume and are not always compatible with a perfect organization. We shall return to this problem of the often antagonistic relationship between organization and functionalization (*cf.* Section 6.7).

6.6.2 Prospects and Challenges Opened up by these Materials

The mesoporous materials, whose origin and first stages of development have just been succinctly described, represent *a very important venture* for the nanosciences. All the solids described are potentially adjustable and functionalizable. Moreover, they are obtained in mild conditions and by methods entirely compatible with any entities (organic, organometallic coordination, biomolecules, particles, etc.) of molecular chemistry.

In fact, these materials can be considered as representing a very good opportunity for the design of interactive materials. Indeed, it is the first time that *chemists have found a general method for introducing in an adjustable solid some molecular units exhibiting properties.* The development of this type of chemistry will permit the precise *localization* both in the pores and in the framework, at the nanometric scale, of functions endowed with useful properties.

The concept of localization of properties is an integral part of material science. However, it could not be developed previously because thermodynamically controlled materials lack synthetic versatility. As we have seen (*cf.* Chapter 3) only hydrothermal synthesis can, in a predictive manner, control the localization and organization of functional entities in a crystalline material. In addition, molecular chemists, working usually in solution, have never shown any interest in the problems of functional localization in solids. The discovery of mesoporous materials *introduces a genuine revolution in the way of thinking for molecular chemists.*

To illustrate this point, let us present schematically a mesoporous material and describe all the possibilities introduced by the controlled localization of functions in the two parts of the material. Figure 6.8 shows schematically a very classical mesoporous material with a hexagonal structure.

Remarkably, this material contains in just a few cubic nanometers a wide range of possibilities. The chemistry which can be developed at present is the following: (1) functionalization on the pores; (2) chemistry and interactions within the pores; (3) introduction of functional entities in the matrix; (4) nature of the inorganic combination constituting the matrix; (5) nature of the spacers.

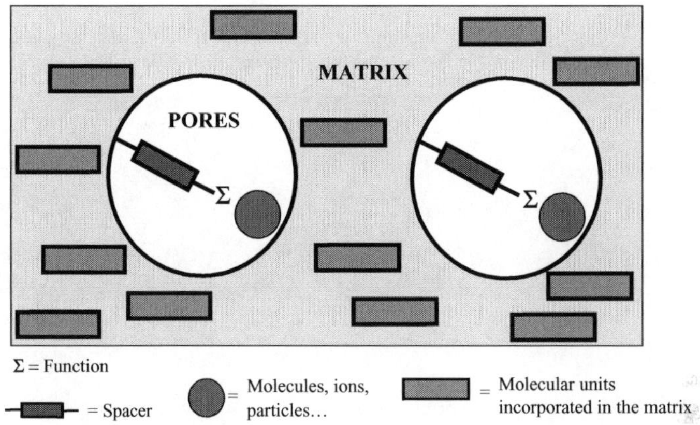

Figure 6.8 Cross-section of a hexagonal mesoporous material showing all possible functionalizations. Reproduced by permission of *L'actualité chimique*

Let us rapidly look through the existing possibilities.

1. **Functionalization of the pores.** This question has already been discussed (*cf.* Section 6.5.2). It is important to stress that the existence of organic entities within the matrix does not influence either the grafting, or the direct synthesis method. These methods permit the introduction of a large variety of functions, as we shall see.
2. **Chemistry and interactions within the pores.** The pores' interior contains a homogeneous (solvent + dissolved compounds) solution. The experimentalist can thus introduce all the necessary chemical products. Chemistry inside the pores, similar to chemistry in solution, can in addition introduce functions on the surface. Moreover, the nanometric size of the pores can induce a confinement effect. This is a most interesting domain for physical and chemical investigations.
3. **Introduction of functional entities in the matrix.** The synthesis of PMOs permits the introduction in SiO_2 matrices of molecular entities having (physical, chemical, etc.) properties. This will be discussed in relation to interactive materials (*cf.* Section 6.8).
4. **Nature of the inorganic combination constituting the matrix.** Using an oxide different from SiO_2 represents an interesting possibility.[48, 49] It is known that many oxides have notable physical properties: SnO_2 (semiconduction), TiO_2 (photovoltaïc effect), NiO, Fe_2O_3, Fe_3O_4 (magnetism), In_2O_3 (electric conduction, photoluminescence, gas detection, etc.). *We can now anticipate multifunctional materials in the near future.* However, the chemical challenge of the sol-gel process for these oxides must be solved.

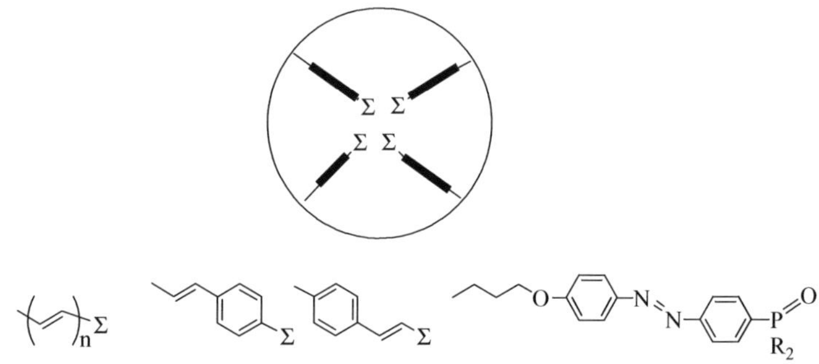

n = 1, 2, 3 $\Sigma = CO_2H, CO_2R, CONHR$ $R = Me, C_6H_5$

Figure 6.9 Some examples of functional spacers which can be introduced into the pores by direct synthesis

5. **Nature of the spacers.** Until now, the spacers have generally been very simple. The entities used are aliphatic $(CH_2)_n$ or aromatic (C_6H_5, $C_6H_5{-}CH_2$,etc.). However, more sophisticated spacers, for example conjugated π systems with possible conducting properties, can be used. Spacers endowed with optical or nonlinear optical properties are also conceivable (Figure 6.9).

Preliminary work has opened up very exciting prospects for organizing matter according to its properties. It is important, however, to point out that most of the chemistry permitting control of the properties of the matrix, of the pores, and of their interactivity has still to be developed. *The basis of this chemistry exists but its development is still to be carried out.*

6.7 IMPORTANCE OF FUNCTIONALIZATION AND OF WEIGHT ANALYSES

Control of the introduction of functions, and of their distribution in pores and in the framework constitutes an indisputable prerequisite for the preparation of mesoporous materials. It is indeed this mastery, which will permit the introduction of (physical, chemical, biological) properties. It represents the *necessary condition* for the development and the exploitation of the extraordinary possibilities of mesoporous materials. However, this condition is not sufficient. As a matter of fact, each property will require that a certain number of supplementary conditions be satisfied in order that these functions truly correspond to a usable property.

In addition, it is important to stress that functionalization and organization are often antagonistic. Indeed, the introduction of chemical functions, of chelating systems, of metal ions or of nanoparticles, will perturb the organization of a hexagonal mesoporous material, since many of them exhibit a non-negligible volume at the nanometric scale. Such systems also modify the charge distribution and the hydrophilicity/hydrophobicity balance. This usually results in the disorganization of the hexagonal system and to the presence of so-called 'vermicular' materials. The X-ray diffractogram is then different from that of a perfect hexagonal system (Figure 6.10).

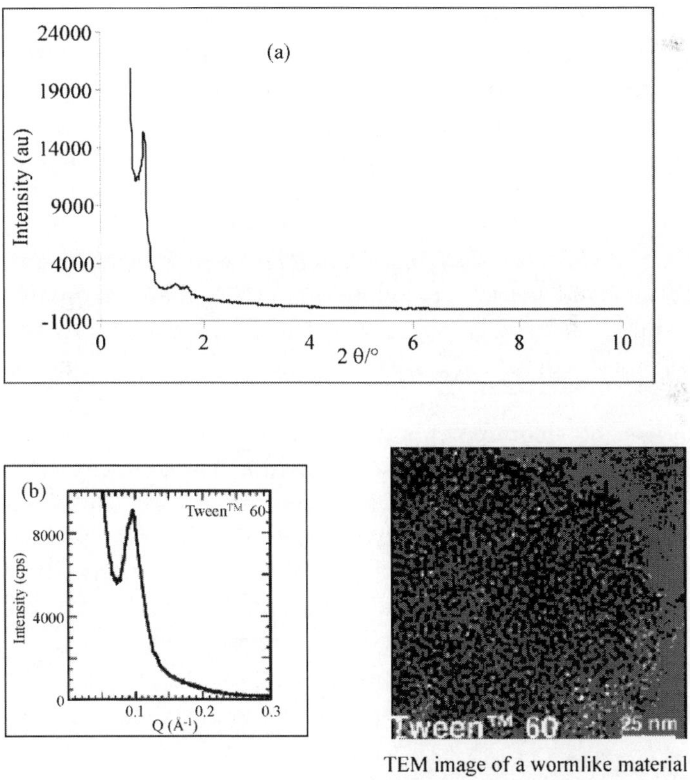

Figure 6.10 (a) X-ray diffractogram of a functionalized hexagonal mesoporous silica. (b) X-ray diffractogram of a functionalized vermicular mesoporous silica and image obtained by transmission electronic microscopy. (a) Reprinted with permission from *Chemical Communications*, Ordered SBA-15 mesoporous silica containing phosphonic acid groups prepared by a direct synthetic approach, Corriu, Robert J. P.; Guari, Yannick; Mehdi, Ahmad; Reye, Catherine; Thieuleux, Chloe; Datas, Lucien, *8*, 763–764. Copyright (2001) RSC. (b) Reprinted with permission from *Chemistry of Materials*, Assembly of mesoporous silica molecular sieves based on nonionic ethoxylated sorbitan esters as structure directors, Prouzet, Eric; Cot, Frederic; Nabias, Georges; Larbot, Andre; Kooyman, Patricia; Pinnavaia, Thomas J., *11*, 1498–503. Copyright (1999) ACS

However, it must be emphasized that the material's properties will always come from functionalization, which therefore must have the first priority. The organization can certainly be improved later by varying the experimental conditions. It is important to point out that a vermicular organization corresponds to a regular arrangement at micrometric scale. It follows that study of the interactions at nanometric scale (correction factor: 10^{-3}) can be justifiably carried out on a vermicular mesoporous material.

Thus, at least in an exploratory phase, functionalization should be favored with respect to organization.

Attention should be drawn to some errors that have appeared in this area of prolific and expanding chemistry of the mesoporous systems. Many authors tend to stress organization, which they consider with reason as necessary for the study and the development of the material properties. However, many of them neglect to do complete analyses, leaving out in particular the centesimal analysis of the elements constituting the material. If physical methods have permitted much progress, *weight analysis remains, however, necessary.* In an entirely new domain and for noncrystalline materials, these analyses give the indispensable quantitative estimates, which are not supplied by either X-ray or microscopy, which can only give an incomplete picture of the material.

Thus, it is not uncommon for materials containing sulfur or nitrogen, to see a description of the material's organization based on X-ray data, while the percentage of N or S has not been determined. Certainly, centesimal analyses cannot be carried out on materials obtained as thin layers by dip or spin coating, as the quantity is too small. *However, analyses of a control sample are compulsory* and must, in addition, be completed by a single or several physical methods. Moreover, Raman study of a thin layer is possible and can be excellent proof of the presence of certain functional groups.

6.8 ON THE WAY TO INTERACTIVE NANOMATERIALS

Interactive nanomaterials have been described as the nanometric version of smart materials, which have two properties capable of interacting with each other. Obviously, mesoporous materials are particularly suited for the preparation of these materials, thanks to the control of functionalization methods. It has been possible to obtain a precise location, either in the pores or in the walls, of chemical entities, which exhibit potential (physical or chemical) properties. In all what follows, the term *functionalization* will correspond to the introduction of a potential property, usually physical, eventually chemical.

A certain number of mesoporous materials satisfying these criteria have been synthesized. These nanomaterials incorporate transition metal ions, lanthanides presenting various (optical, magnetic, etc.) properties,[45] or organic molecules having optical or nonlinear optical properties.[50] These entities have been attached either to the pores or to the framework of the material.

6.8.1 Examples of Joint Functionalization of the Framework and the Pores

A tetra-azamacrocycle (cyclam) suitably substituted on the four nitrogens by groups bearing $Si(OR)_3$ functions has been introduced in the framework of a mesoporous material in the presence of the block copolymer P123 $[H-(OCH_2CH_2)_{20}-(OCH(CH_3)-CH_2)_{70}-(O-CH_2CH_2)_{20}OH]$ (Figure 6.11). This type of material has permitted the controlled localization of metallic ions either in the framework or in the pores. The simultaneous introduction of this type of ions in both places is done by direct chelation in the matrix followed by the grafting of substituted

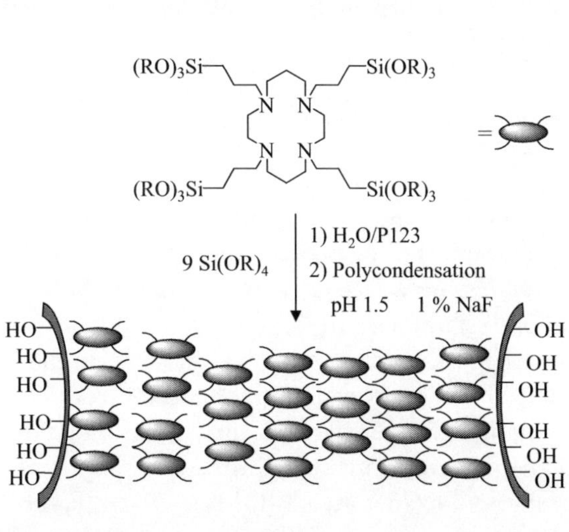

Figure 6.11 Introduction of a tetra-azamacrocycle into the framework of a mesoporous silica

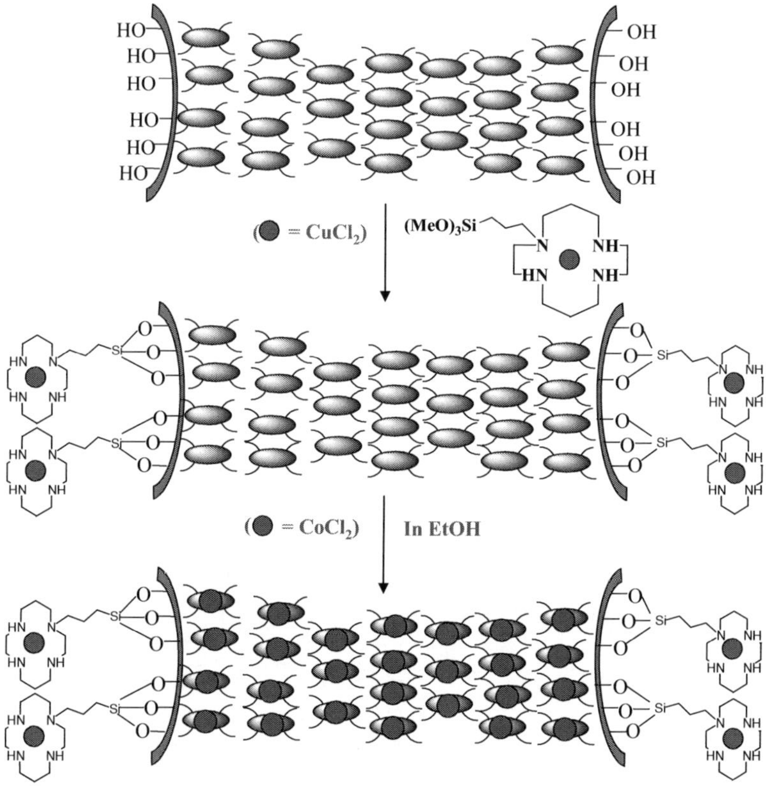

Figure 6.12 Introduction of transition metal ions (Co^{2+}, Cu^{2+}) either in the walls or in the pores of the mesoporous material shown in Figure 6.11. Reprinted with permission from *New Journal of Chemistry*, Control of coordination chemistry in both the framework and the pore channels of mesoporous hybrid materials, Corriu, Robert J. P.; Mehdi, Ahmad; Reye, Catherine; Thieuleux, Chloe, 27, 905–908. Copyright (2003) RSC (See Plate 8 for color representation)

cyclams on the pores. The accessibility of the ions in the matrix is remarkable and deserves to be pointed out. *In all cases, transition metal ions and lanthanides can be directly incorporated into the solid matrix.* Each of these ions exhibits a different physical property (Figure 6.12).

It has also been possible to introduce lanthanides in the walls by direct chelation on material bearing cyclams in the framework (Figure 6.13) at a ratio of one lanthanide ion for two cyclams, as described in Section 5.7.3. Transition metal ions can be introduced directly in the pores using chelating cyclam units functionalized with a $Si(OMe)_3$ group (Figure 6.13).

Thus, ions with quite diverse properties (Eu^{3+}, Er^{3+}, Tb^{3+}, photoluminescence; Cu^{2+}, paramagnetism; Gd^{3+}, Co^{2+}, Ni^{2+}, Fe^{3+}, magnetism)

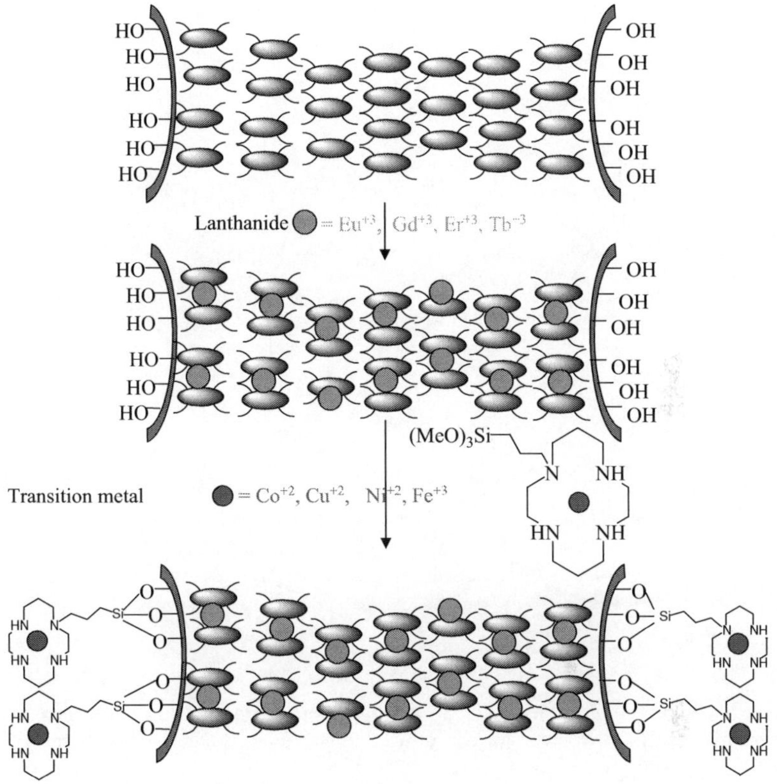

Figure 6.13 Introduction of lanthanides in the framework (one ion for two chelating units) and fixation of transition metals in the pores (See Plate 9 for color representation)

can be introduced in the matrix and in the pores. These materials represent potential interactive nanomaterials since two types of ions with different properties can be precisely located at the nanometric scale.

Let us now present two examples in which other properties (photoluminescence, electroluminescence and nonlinear optical) have been introduced.

In the first case, direct synthesis can introduce Er^{3+} ions chelated by three phosphine oxides in the matrix.[51] These complexes have IR photoluminescent properties. In the course of the same synthesis, monosilylated spacers with nonlinear optical properties are attached to the pores (Figure 6.14). Their phosphine oxide function allows later fixation of metal ions with physical properties.

In the second example, the material is again obtained by direct synthesis: a molecule with nonlinear optical properties is introduced in the

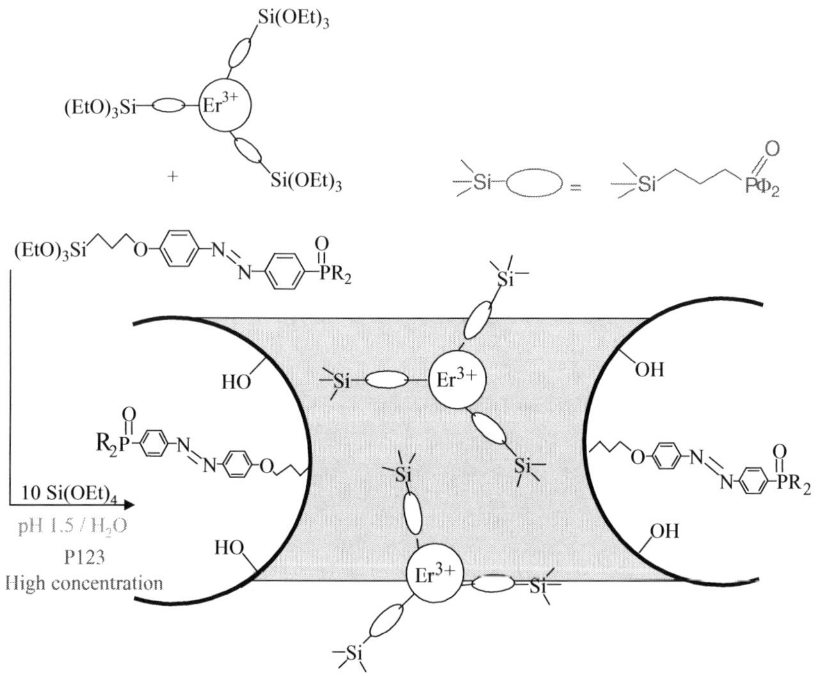

Figure 6.14 One-step synthesis of a mesoporous material bearing a complex having photoluminescent properties in the matrix and a spacer endowed with nonlinear optical properties and able to chelate metallic ions

matrix, while $-(CH_2)_3SH$ groups are fixed to the pores. It is then possible to incorporate gold nanoparticles (Figure 6.15).[50]

These results show that it is possible to position very precisely chemical functions having different potential physical properties either in the walls or in the pores. Thus, investigation on interactive nanomaterials is not pure utopia. Chemists already have tools for preparing nanomaterials in which two different properties can interact at the nanometric scale.

Continuation of this topic can only be done in close collaboration with physicists, at the *conception* as well as at the *realization* and *investigation* levels.

6.8.2 An Acid and a Base at the Nanometric Scale

The possibilities offered by interactive nanomaterials are huge. In order to illustrate the high number of synthetic possibilities, a material in which an acidic function is introduced in the framework and a basic function in the

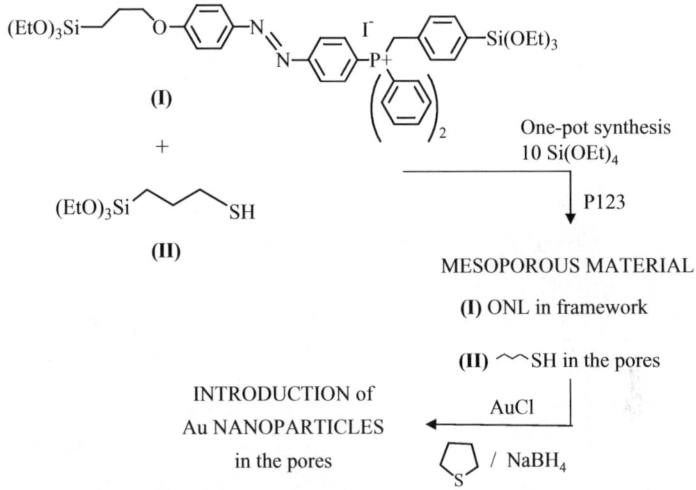

Figure 6.15 Two-step synthesis of a material with nonlinear optical properties in the walls and bearing gold nanoparticles in the pores

pores has been prepared. The two functions are located at nanometric distances from one another and preserve their properties. This material is synthesized according to the scheme shown in Figure 6.16.[52]

The material is obtained by direct synthesis using a α,ω-bis (trimethoxysilyl)alkyldisulfide in the presence of the triblock copolymer P123 and one amine Boc-protected (Figure 6.16). In one step, the disulfide is introduced into the framework and the protected amine in the pores. The acid function is obtained by a sequence of reactions in the framework: reduction of the S–S bond by $NaBH_4$, followed by oxidation of the SH group into SO_3H. The amine is then freed by heating (160 °C in vacuum). All these reactions occur with yields higher than 85%. At that stage, the material has acidic functions in the walls and basic functions in the pores. Their availability is shown by their reactions. The acid reacts quantitatively with Li_2CO_3 while the amine function reacts with acrylamide in a Michael reaction. Note that the system behaves rather like a membrane since the yields of the reactions occurring inside the pores depend on the nature of the solvent. These results were obtained in THF. However, when the Michael reaction is carried out in protic solvents (EtOH and H_2O), a dramatic drop in yields is observed (2% in H_2O and 10% in EtOH). (The yield is $>85\%$ in THF.) The reason is that protic solvents permit proton exchange between the acidic functions in the framework and the basic functions in the pores. In contrast, non-protic solvents (THF, CH_3CN), which do

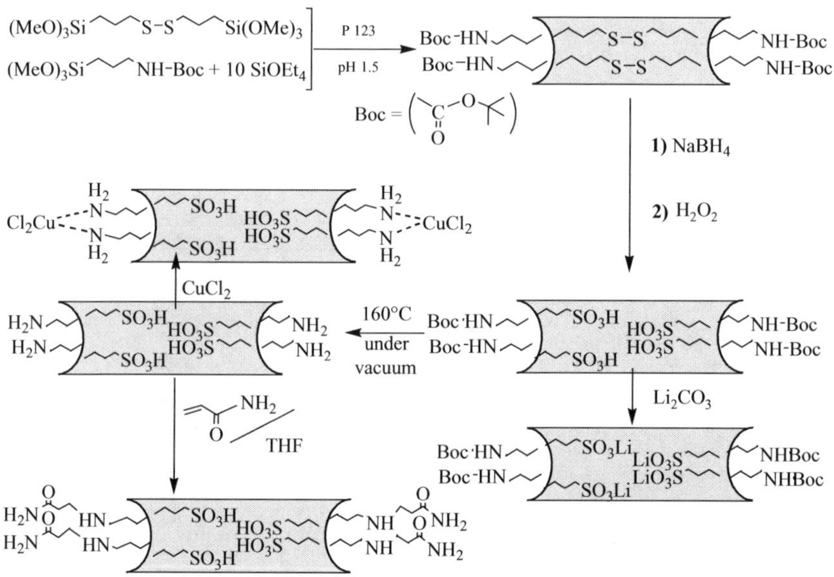

Figure 6.16 Reaction scheme leading to a bifunctional nanomaterial (acidic in the framework and basic in the pores). Reprinted with permission from *Journal of the American Chemical Society*, Mesoporous materials with an acidic framework and basic pores. a successful cohabitation, Alauzun, Johan; Mehdi, Ahmad; Reye, Catherine; Corriu, Robert J. P., *128*, 8718–8719. Copyright (2006) ACS

not allow proton transfer, permit the cohabitation of the two antagonistic functions at the nanometric scale.

6.9 PREPARATION OF NEW MATRICES

From current knowledge, we can list the properties capable of producing interesting developments.

First, let us consider the properties of the matrix. As already mentioned (Sections 3.6.5.2 and 5.7.4.3), some matrices different from SiO_2 can bring in intrinsic physical properties. Interesting but limited results have been obtained with TiO_2 and Al_2O_3. The development of inorganic polymerization of the sol-gel type for magnetic oxides (such as Fe_2O_3, Fe_3O_4 and NiO) can be anticipated.

In the case of silica matrices – whose polymerization and shaping technology are well mastered – the physical properties due to the entities incorporated in the framework constitute a domain still little explored. Initial results suggest that the inclusion of nano-objects will lead to unexpected properties. This is just an 'appetizer' and we can foresee a considerable development of this type of materials.

Functionalization of the pores is in full expansion but many unknowns remain. Physical properties of mesoporous materials containing metal ions (or metal oxides) have been the subject of only a very limited number of studies. Insertion of metal nanoparticles in the pores of nonfunctionalized mesoporous silica has been much studied. In many cases, the introduction in the pores is irregular and the sizes of particles are quite diverse. This route gives an unsatisfactory control of the distribution and the size of the nanoparticles introduced in the pores, except for small quantities. Only an appropriate functionalization can produce new materials endowed with a precise property. Polyfunctionalization inside the pores has not been studied extensively.

The existence of a confinement effect in the pores has been mentioned but the results obtained are scarce. Study of the conductivity of the ions inside the pores is only now beginning to be considered.

The importance of spacers capable of establishing an interaction between the properties of the matrix and those of the pores remains to be explored. This list is far from exhaustive. It shows, however, the extent of possibilities offered to researchers working in this field.

6.10 ON THE WAY TO BIOLOGICAL APPLICATIONS

The functionalization of pores of mesoporous materials opens up very interesting prospects in the biological field. Introduction of an alcohol or amine function permits the fixation of drugs by covalent bonds which could be selectively cleaved in precise conditions. These materials then come as potential agents for delivering drugs *in situ* and in a controlled manner. Thus, ibuprofen, a non-steroidic anti-inflammatory agent, has been grafted on the pores of a mesoporous silica[53] (cf. Figure 6.17). The aim is the regular *in situ* delivery of a drug by controlled hydrolysis of an ester function.

Mesoporous systems have been widely used to immobilize biological species. Enzymes[54–56] have been encapsulated in silica obtained by the sol-gel process. This immobilization makes them more active since they are in an environment whose hydrophilic/hydrophobic balance is appropriate for an optimal efficiency. These supported enzymes are selective biocatalysts for many interesting transformations in organic chemistry.

Recently, lipase,[57] which permits selective hydrolysis of esters and thioesters, has been encapsulated in several mesoporous silica with different characteristics. The results obtained are most promising.

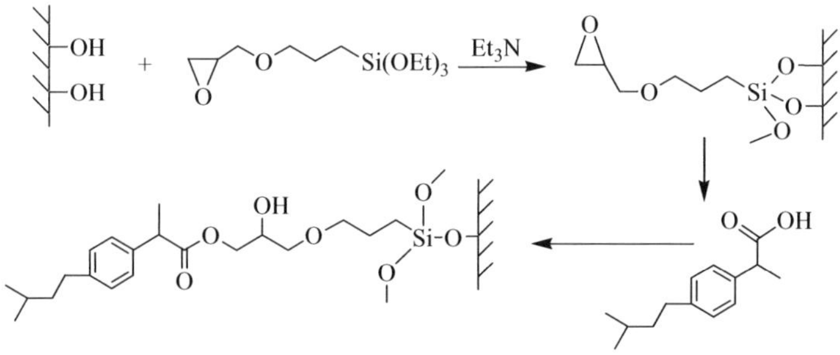

Figure 6.17 Introduction of a drug inside the pores of a mesoporous silica. Reprinted with permission from *New Journal of Chemistry*, Synthesis and characterization of ibuprofen-anchored MCM-41 silica and silica gel, Tourne-Peteilh, Corine; Brunel, Daniel; Begu, Sylvie; Chiche, Bich; Fajula, Francois; Lerner, Dan A.; Devoisselle, Jean-Marie, *27*, 1415–1418. Copyright (2003) RSC

6.11 CONCLUSIONS

This chapter showed the possibilities opened up by mesoporous systems. As we have seen in Section 6.6.2, it is the first time that molecular chemists have had the chance to think of their chemistry in terms of precise localization and interactivity. Considering Figure 6.8 which introduces the work done so far, it is possible to rapidly bring to mind the potential developments of this type of materials (Figure 6.18).

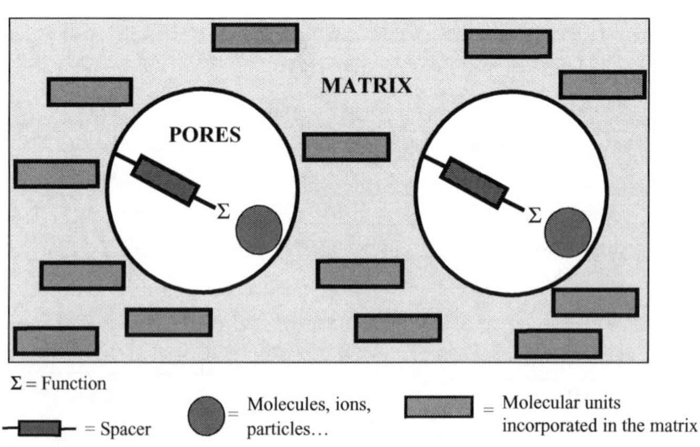

Figure 6.18 Cross-section of a mesoporous material with hexagonal structure presenting all the possibilities of functionalization. Reproduced by permission of *L'actualité chimique*

We have described some developments regarding interactive materials, limiting ourselves to materials already obtained and suggesting some extensions. This is only a small part of the existing possibilities.

It is the synthetic know-how which will permit the preparation of *matrices endowed with physical (magnetic, electric, optical, etc.) properties* by varying the incorporation of nano-objects and also the nature of the oxide. We have given some examples of metal ions and organic entities capable of bringing physical properties into the material framework. A vast domain awaits the investigators.

The *distribution of the functions inside the pores* is also an important consideration. Pores capable of coupling two functions deserve to be studied. Published works give little information on the distribution of two functions located in the pores of a mesoporous material. However, they all point to a very fruitful domain for experimentation.[58–61]

Clearly, considering that the framework and pores are functionalized in different manners, the problem of the interaction between these properties arises. In particular, it may be opportune to examine the *role of the spacer*. For example, a conjugated spacer will be a conductor while a saturated spacer will be an insulator.

Chemists interested in this field must take into account all of the possible interactions because they are all nanometric.

REFERENCES

[1] C.T. Kresge, M.E. Leonowicz, W.J. Roth, J.C. Vartuli, J.S. Beck, *Nature* **1992**, *359*, 710.

[2] J.S. Beck, J.C. Vartuli, W.J. Roth, M.E. Leonowicz, C.T. Kresge, K.D. Schmitt, T.-W. Chu, D.H. Olson, E.W. Sheppard, S.B. McCullen, J.B. Higgins, J.L. Schlenker, *J. Am. Chem. Soc.* **1992**, *114*, 10834.

[3] T. Yanagisawa, T. Shimizu, K. Kuroda, C. Kato, *Bull. Chem. Soc. Jpn* **1990**, *63*, 988.

[4] P.T. Tanev, T.J. Pinnavaia, *Science* **1995**, *267*, 865.

[5] P.T. Tanev, T.J. Pinnavaia, *Chem. Mater.* **1996**, *8*, 2068.

[6] E. Prouzet, T.J. Pinnavaia, *Angew. Chem., Int. Ed. Engl.* **1997**, *36*, 516.

[7] S.A. Bagshaw, E. Prouzet, T.J. Pinnavaia, *Science* **1995**, *267*, 1242.

[8] D. Zhao, Q. Huo, J. Feng, B.F. Chmelka, G.D. Stucky, *J. Am. Chem. Soc.* **1998**, *120*, 6024.

[9] A. Walcarius, C. Delacôte, *Chem. Mater.* **2003**, *15*, 4181.

[10] D.J. Macquarrie, D.B. Jackson, J.E.J. Mdoe, J.H. Clark, *New J. Chem.* **1999**, *23*, 539.

[11] C.J. Brinker, Y. Lu, A. Sellinger, H. Fan, *Adv. Mater.* **1999**, *11*, 579.

[12] A. Gibaud, J.F. Bardeau, M. Dutreilh-Colas, M. Bellour, V.V. Balasubramanian, A. Robert, A. Mehdi, C. Reye, R.J. Corriu, *J. Mater. Chem.* **2004**, *14*, 1854.

[13] S. Dourdain, J.-F. Bardeau, M. Colas, B. Smarsly, A. Mehdi, B.M. Ocko, A. Gibaud, *Appl. Phys. Lett.* **2005**, *86*, 113108/1.

[14] A. Mehdi, S. Dourdain, J.-F. Bardeau, C. Reye, R.J.P. Corriu, A. Gibaud, *J. Nanosci. Nanotechnol.* **2006**, *6*, 377.
[15] P. Schmidt-Winkel, W.W. Lukens, D. Zhao, P. Yang, B.F. Chmelka, G.D. Stucky, *J. Am. Chem. Soc.* **1999**, *121*, 254.
[16] N. Huesing, C. Raab, V. Torma, A. Roig, H. Peterlik, *Chem. Mater.* **2003**, *15*, 2690.
[17] D. Brandhuber, V. Torma, C. Raab, H. Peterlik, A. Kulak, N. Huesing, *Chem. Mater.* **2005**, *17*, 42621.
[18] W. Zhang, M. Froba, J. Wang, P.T. Tanev, J. Wong, T.J. Pinnavaia, *J. Am. Chem. Soc.* **1996**, *118*, 9167.
[19] L. Mercier, T.J. Pinnavaia, *Adv. Mater.* **1997**, *9*, 500.
[20] A. Cauvel, G. Renard, D. Brunel, *J. Org. Chem.* **1997**, *62*, 749.
[21] P.M. Price, J.H. Clark, D.J. Macquarrie, *J. Chem. Soc., Dalton Trans.* **2001**, 101.
[22] S.L. Burkett, S.D. Sims, S. Mann, *Chem. Commun.* **1996**, 1367.
[23] D.J. Macquarrie, *Chem. Commun.* **1996**, 1961.
[24] D.J. Macquarrie, D.B. Jackson, S. Tailland, K.A. Utting, *J. Mater. Chem.* **2001**, *11*, 1843.
[25] Y. Mori, T.J. Pinnavaia, *Chem. Mater.* **2001**, *13*, 2173.
[26] R.J.P. Corriu, C. Hoarau, A. Mehdi, C. Reye, *Chem. Commun.* **2000**, 71.
[27] R.J.P. Corriu, F. Embert, Y. Guari, A. Mehdi, C. Reye, *Chem. Commun.* **2001**, 1116.
[28] D. Margolese, J.A. Melero, S.C. Christiansen, B.F. Chmelka, G.D. Stucky, *Chem. Mater.* **2000**, *12*, 2448.
[29] C.E. Fowler, S.L. Burkett, S. Mann, *Chem. Commun.* **2004**, 1742.
[30] T. Yokoi, H. Yoshitaki, T. Tatsumi, *Chem. Mater.* **2003**, *15*, 4536.
[31] A. Mehdi, C. Reye, S. Brandes, R. Guilard, R.J.P. Corriu, *New J. Chem.* **2005**, *29*, 965.
[32] S. Inagaki, S. Guan, Y. Fukushima, T. Oshuna, O. Terasaki, *J. Am. Chem. Soc.* **1999**, *121*, 9611.
[33] B.J. Melde, B.T. Holland, C.F. Blanford, A. Stein, *Chem. Mater.* **1999**, *11*, 3302.
[34] T. Asefa, M.J. MacLachlan, N. Coombs, G.A. Ozin, *Nature* **1999**, *402*, 867.
[35] C. Yoshina-Ishii, T. Asefa, N. Coombs, M.J. MacLachlan, G.A. Ozin, *Chem. Commun.* **1999**, 2539.
[36] T. Asefa, M.J. MacLachlan, H. Grondey, N. Coombs, G.A. Ozin, *Angew. Chem., Int. Ed.* **2000**, *39*, 1808.
[37] G. Temtsin, T. Asefa, S. Bittner, G.A. Ozin, *J. Mater. Chem.* **2001**, *11*, 3202.
[38] S. Inagaki, S. Guan, T. Oshuna, O. Terasaki, *Nature* **2002**, *416*, 304.
[39] Y. Goto, S. Inagaki, *Chem. Commun.* **2002**, 2410.
[40] J. Morell, G. Wolter, M. Fröba, *Chem. Mater.* **2005**, *17*, 804.
[41] M. Alvaro, B. Ferrer, V. Fornés, H. Garcia, *Chem. Commun.* **2001**, 2546.
[42] C. Baleizao, B. Gigante, D. Das, M. Alvaro, H. Garcia, A. Corma, *Chem. Commun.* 2003, 1860.
[43] M. Alvaro, M. Benitez, D. Das, B. Ferrer, H. Garcia, *Chem. Mater.* **2004**, *16*, 2222.
[44] R.J.P. Corriu, A. Mehdi, C. Reye, C. Thieuleux, *Chem. Commun.* **2002**, 1382.
[45] R.J.P. Corriu, A. Mehdi, C. Reye, C. Thieuleux, *New J. Chem.* **2003**, *27*, 905.
[46] O. Olkhovyk, M. Jaroniec, *J. Am. Chem. Soc.* **2005**, *60*, 127.
[47] E. Besson, A. Mehdi, D.A. Lerner, C. Reye, R.J.P. Corriu, *J. Mater. Chem.* **2005**, *15*, 803.
[48] F. Schüth, *Chem. Mater.* **2001**, *13*, 3184.
[49] X. Hu, D. Antonelli, *Angew. Chem., Int. Ed.* **2002**, *41*, 214.

[50] E. Besson, A. Mehdi, V. Matsura, Y. Guari, C. Reye, R.J.P. Corriu, *Chem. Commun.* **2005**, 1775.
[51] E. Besson, A. Mehdi, C. Reye, R.J.P. Corriu, *J. Mater. Chem.* **2006**, *16*, 246.
[52] J. Alauzun, A. Mehdi, C. Reye, R.J.P. Corriu, *J. Am. Chem. Soc.* **2006**, *128*, 8718.
[53] C. Tourné-Peteilh, D. Brunel, S. Bégu, B. Chiche, F. Fajula, D.A. Lerner, J.-M. Devoiselle, *New J. Chem.* **2003**, *27*, 1415.
[54] M.T. Reetz, J. Simpelkamp, A. Zonta, German Patent DE 4408152 Al, **1995**.
[55] M.T. Reetz, *Adv. Mater.* **1997**, *9*, 943.
[56] Y.-J. Han, G.D. Stucky, A. Butler, *J. Am. Chem. Soc.* **1999**, *42*, 121.
[57] A. Galarneau, M. Mureseanu, S. Atger, G. Renard, F. Fajula, *New J Chem.* **2006**, *30*, 1.
[58] S.R. Hall, C.E. Fowler, B. Lebeau, S. Mann, *Chem. Commun.* **1999**, 201.
[59] D.J. Macquarrie, *Green Chem.* **1999**, 195.
[60] S. Huh, H.-T. Chen, J.W. Wiench, M. Priski, V.S.-Y. Lin, *J. Am. Chem. Soc.* **2004**, *126*, 1010.
[61] R. Mouawia, A. Mehdi, C. Reye, R.J.P. Corriu, *New J. Chem.* **2006**, *30*, 1077.

7

Prospects and Stakes

7.1 GENERAL REMARKS

In sciences, development forecasts are difficult, as they will unavoidably be incorrect. This is true quantitatively, for overestimation and underestimation are unavoidable, as they depend on a number of unknown parameters. This is also true qualitatively, as the advances in fundamental research lead to the discovery of new possibilities, capable of leading to interesting prospects but also to some hurdles difficult to overcome.

The economical stakes are important: all developed countries have created interdisciplinary programs on '*nanosciences and nanotechnologies*'. It is obvious that the *materials of the future will not simply be an extension of existing materials and systems. They must be capable of coupling several functions in an interactive manner.* We shall witness a complete upheaval of technologies, due to the discovery and use of methods based on entirely new principles.

The prospects opened up are far reaching. It is important to stress that *electronics* is not the only domain concerned with nanosciences, even if its place is indisputable and of major interest. This is the consequence of its present development and of the economical stakes it represents. Mostly, the top-down approach is favored, consisting in a miniaturization of existing materials using new technologies. The currently stated objective is to reach a size of $<0.05\,\mu m$. Chemistry will intervene in its customary role as a purveyor of special plastic compounds, of new additives, etc. Its role will in no way be determining in this context.

Molecular Chemistry of Sol-Gel Derived Nanomaterials Robert Corriu and Nguyên Trong Anh

The same is not true with the bottom-up approach, which, for the moment, is limited to the exploration of the nanoworld. This exploration reveals properties very different from those at the micrometric scale. The carbon nanotubes case is particularly illustrative. These chemical entities are produced during incomplete oxidation of carbon. They are made of nonplanar $>C=C<$ double bonds lying on curved surfaces, which modify the properties of π electrons of the unsaturated carbons. Nanotubes present special electronic properties. For example, they can behave like electronic components. For now, the participation of chemistry is of a fundamental nature, since the domain is only being explored. However, in the future, it will be useful for perfecting the control of syntheses and of appropriate doping agents.

It must be noted, however, that the development of nano-objects and their confinement in suitable matrices will permit the emergence of unusual chemical behaviors as well as the discovery of new physical properties related to the scale change and the coupling of properties. These novelties will lead inevitably to new domains in which chemistry will find its place, on account of its power of creation, of adaptation and of organization of matter. However, these domains must be interactively exploited with other scientists, in particular physicists and biologists. This interaction implies communal thinking, beginning at the conception and continuing throughout the realization.

We shall now present some developments induced by nanosciences. The examples, which will be succinctly discussed, are simple extrapolations of promising research currently in progress. However, they will give the reader a notion of the potential stakes. It is clear that many objectives have certainly been overlooked.

7.2 PREDICTABLE DEVELOPMENTS

Under the term 'molecular photonics'[1] are regrouped two important domains: *optical telecommunications*; and *storage and treatment of information*.

Storage and treatment of information stem from electronics and, until now, are limited to the two dimensions of the surface of silicon. The possibilities offered by nanomaterials for magnetic or optical storage appear, a priori, much greater, since they use the three dimensions of space.

Optical storage is an extrapolation of the confocal methods already used in biological microscopy, which can explore a volume with a separating power of the order of a micrometer. Two-photon confocal methods have recently attained a precision of a tenth of a micrometer,[2] an additional order of magnitude with respect to the hopes of electronics. Optical methods open up interesting prospects as more parameters, such as the intensity, the wavelength and the polarization of the beams, can come into play. In the longer term, new materials can permit information storage beyond binary languages, as a result of the selective polarization of different centers. This prospect, inconceivable in silicon-based electronics, is theoretically possible in optical systems.

Optical telecommunications represent an important advance with respect to existing electronic methods, which are slower and less precise. This domain implies a close cooperation between chemists and physicists since it concerns more or less all subfields of chemistry. For example, some organic molecules have interesting optical properties. Transition metals and lanthanides all exhibit optical properties, which vary according to their coordination state, *i.e.* depending on the chelating organic entities. On account of this, research and development in telecommunications imply, in many cases, the invention of completely new materials. Coordination chemistry, molecular chemistry and material chemistry are sufficiently mature to permit the preparation and shaping of nanomaterials organized and optimized at different scales, from the nanometric to micrometric (and even beyond). In addition, these nanomaterials can have adjustable optical properties. The research fields opened up by nonlinear optics appear to be very promising.

Research studies on magnetic methods for information storage and, in particular, on nanomaterials having magnetic properties are much less advanced. This topic is yet to be developed to offer ways for designing new magnetic materials.

Highly selective *sensors* are implicated in many domains, ranging from environmental problems to biomedical applications. This field of research should grow in the next few years. As a matter of fact, electronics can very satisfactorily sense, amplify, treat and quantify low intensity signals. For their part, chemists have acquired an excellent knowledge of the chelation modes of ions and inorganic particles. The understanding of interactions is making great progress. In addition, systems, which can interact with biological molecules, are better and better identified, at least in a number of cases. Microsensors capable of detecting and measuring *in situ* the precise amount of specific biological substances are crucial in the health sector. They will become operative as soon as semiconductors can be directly functionalized by a chemical graft capable of transforming a selective interaction into an electronic signal. Note that detection and identification of pollutants constitute also a major problem for society.

Applications concerning *solar energy* are very important, since energy demand is increasing. Silicon is the most frequently used. However, the energetic cost of its production is enormous ($\sim$8700 kcal kg^{-1}). In addition, its preparation entails an electrochemical process using graphite electrodes. This is certainly not the best choice. The yield of the other cells is at present very low: the photovoltaic yield of TiO_2, the most used sensor, is <10%. The reason is that TiO_2 captures only a weak part of the solar spectrum, which, moreover, is located in the blue and violet range. Any new materials permitting a yield increase represent a crucial development. The solutions, on the one hand, appear to lie in the use of doping agents permitting a better transformation of light energy into electric energy. On the other hand, the materials should be shaped and completed by additional optical devices in order to recover a maximum amount of solar energy. The societal goal of the optimal recovery of solar energy is of basic importance. It is clear that inorganic polymerization opens up more possibilities for the creation of new materials combining various types of (organic and inorganic) chemical entities in various geometries. These materials with efficient properties will most probably be found in the wide range of possibilities opened up by the combinations of oxides, nitrides and phosphides and sulfides.

Some other nanoscience domains, such as *catalysis, selective separations* and *heterogeneous chemical synthesis* belong more specifically to chemistry, although they also have recourse to physics and technology.

Heterogeneous catalysis has a considerable economic importance: the major processes of petrochemistry, of polymer and specialty products synthesis all make use of supported catalysts. This discipline is quite suitable for a nanometric approach. The objective of chemists will consist of optimizing the catalyst's efficiency using a material conceived and synthesized for that purpose. This accomplishment implies on the one hand a perfect understanding of the catalytic act and on the other hand, the mastery of the control of the introduction and distribution of functional groups on the surface of the material. In addition, the material should permit the circulation of the reagents and the recovery of the products.

For such an approach, mesoporous materials appear *a priori* to be excellent candidates. It is indeed possible to control the pore size and the distribution of functionalities inside the pores. It is also possible to introduce a very large range of functional groups either by grafting or by direct synthesis. Many different functionalities can thus be fixed on the walls of the pores. The geometry of the catalytic process can therefore be modified. As a matter of fact, since the mechanism of the catalytic process and its geometry are determined from *homogeneous* catalysis data, an adaptation will probably be necessary to optimize heterogeneous catalysis. This nanometric approach would thus give catalysts 'tailor-made' from mesoporous materials.

Enzymes have been immobilized on mesoporous systems. They represent an efficient and very selective catalysis, since they can favor a specific reaction on a particular substrate.

In addition, the nanometric approach will provide catalysis with new tools, which cannot be obtained by existing methods.

Let us mention for example *polyfunctional catalysts* capable of bringing about several transformations in a selective manner. This vision will come true when chemists know how to organize different chemical functions within the same material. This prospect is not utopian, since control and regular distribution of functions within the pores of an organized mesoporous system have already been accomplished.[3] *The physical properties of the matrix can also be used to amend the catalytic process*. For example, a TiO_2 matrix can modify electrically the catalytic site and a magnetic nanomaterial can orient certain molecules in a field.

As far as the applications and the economic and social problems are concerned, *selective separation* is certainly one of the most promising topics. Selective depollution of organic and inorganic substances can be accomplished by nanomaterials bearing chemical entities capable of separating undesirable (organic, inorganic, biological) substances by selectively chelating or complexing them. For example, such materials can solve the problem of nuclear waste. Indeed, in the near future, a new process[4] will be able to simplify the effluent decontamination process (*cf.* Chapter 4). As a result of a selective chelation of uranides, the radioactivity of the waste will be at the limits of detection ($5\,Bq\,M^{-3}$), largely lower than the natural radioactivity.

A selective and efficient chemical separation is the necessary condition for the complete reprocessing of nuclear combustibles, as well as for their storage in smaller and smaller volumes. Among other important applications, let us mention the chemical methods of isotopic separation[5] (*cf.* Section 4.5), which are as efficient but much less expensive than the methods used now. Moreover, it is important to point out that applications are not limited to nuclear energy, since the physical and chemical properties of isotopes are not strictly identical. The replacement of one isotope for another is susceptible to modification of some of the properties of materials.*

Treatments of water and effluents, as well as gas purification are essential for the environment. The state of the art is described in Section 4.5. The chelating entities are bonded to polymer resins or to the surface of SiO_2. New developments can be expected since with mesoporous nanomaterials, a regular distribution of grafted functions on a large and totally accessible surface is possible. The examples given previously (*cf.* Section 4.5) concern the selective separations of ions, but mesoporous materials, whose pores can be variously functionalized (*cf.* Chapter 6), can also be used for the separation of organic molecules or any other chemical entities. In addition, the hydrophilic/lipophilic balance of these materials can be modified by chemical treatment of the SiOH functions on the surface. This has not been studied in an exhaustive manner, but the modification, even the optimization of the hydrophilicity of pores by grafting organic entities to the surface is certainly realizable. Mesoporous nanomaterials, owing to their chemical versatility, are, therefore, very efficient tools for selective depollution. Their great specific surface makes their storage ability very important.

Heterogeneous chemical synthesis can afford great modifications in chemical processes. This topic has known a certain success and the recent progress obtained with organic-inorganic hybrids opens up great prospects. Let us recall that the recovery and the elimination of solvents are vital for the chemical industry. In addition, to protect the health of workers, use of certain (benzenic, chlorinated, etc.) solvents is severely restricted. Heterogeneous synthesis could be the answer to several of these difficult problems.

Some chemists[6] have realized one-pot cascade reactions using a heterogeneous medium containing silica particles grafted with functions capable of activating or catalyzing different reactions. For this purpose, they have prepared SiO_2 particles on which they have grafted the appropriate catalysts. The crucial point is that no chemical reaction occurs between the different particles. In fact, the hybrid material has been synthesized in such a way that the chemical functionalization takes place essentially inside a material having a high internal porosity (*cf.* Section 4.7).

---

* The size of the isotopes being different, their electrons do not exhibit the same energy. The possible reactions and properties are very similar but not identical.

The chemical compounds to be transformed remain in solution. They travel from one particle to another undergoing the planned cascade of reactions. Obviously, the cascades must be carefully studied to optimize the efficiency of the process and to avoid undesired reactions. This methodology is particularly promising as it avoids the separation of the products formed in the intermediate steps. For the moment, only a limited number of studies have been carried out. However, this methodology has a bright industrial future because of the high cost of separations and of the possible automatization of this type of process.

The possibilities opened up by the nanosciences in the *biological and medical field* will certainly be of considerable importance too. Miniaturization and specificity of sensors will permit exceptional developments in the health practice. In particular, *in situ* perfectly targeted disease assessment and disease treatment will be possible. Biochip studies are soaring. These chips can detect and dose substances *in vivo*, especially in blood. This detection, by triggering a device incorporated *in situ*, allows the automatic delivery of the appropriate drug. The chip is made up of a surface having functionalities capable of detecting biological entities (sugar in blood in the case of diabetes, for example) by optical recognition with a control cell. The latter will trigger the device liberating the implanted drug once the minimal quantity is reached.

Drug delivery will be completely modified. The controlled delivering *in situ* of drugs will become possible by the development of appropriate nanomaterials. This approach will be also used in agronomy for perfecting devices, which can release the appropriate quantities of pesticides, fungicides or fertilizers at the required time and in a rigorously controlled manner.

REFERENCES

[1] J. Zyss, (ed), *C. R. Acad. Sci.* **2002**, *3*, (4).

[2] S. Kawata, H.B. Sun, T. Tianaka, K. Takada, *Nature* **2001**, *412*, 697.

[3] R.J.P. Corriu, F. Embert, Y. Guari, A. Mehdi, C. Réyé, *Chem. Commun.* **2001**, 1116.

[4] (a) F. Barbette, F. Rascalou, H. Chollet, J.-L. Babouhot, F. Denat, R. Guilard, *Anal. Chim. Acta* **2004**, *502*, 179; (b) R. Guilard, H. Chollet, P. Guiberteau, P. Cocolios, WO 96 29363, **1996**; (c) H. Chollet, J.-L. Babouhot, F. Barbette, R. Guilard, WO 01 15806, **2001**.

[5] (a) M. Lemaire, J. Foos, A. Guy, F. Chitry, S. Pellet-Rostaing, Vigneau O.WO 20022053269, **2002**; (b) F. Chitry, S. Pellet-Rostaing, O. Vigneau, M. Lemaire, *Chem. Lett.* **2001**, 770.

[6] (a) F. Gelman, J. Blum, D. Avnir, *New J. Chem.* **2003**, *27*,205; (b) F. Gelman, J. Blum, D. Avnir, *Angew. Chem., Int. Ed.* **2001**, *40*, 3647.

Index